BK 629.44 G619S
SPACE COMMERCE : FREE ENTERPRISE ON THE HIGH
FRONTIER /GOLDMAN, N
C1985 25.00 FV

3000 696004 30018
St. Louis Community College

WITHDRAWN

```
629.44 G619s                              FV
GOLDMAN
  SPACE COMMERCE : FREE ENTER-
  PRISE ON THE HIGH FRONTIER
                                       25.00
```

St. Louis Community College

Library

5801 Wilson Avenue
St. Louis, Missouri 63110

SPACE COMMERCE

SPACE COMMERCE
Free Enterprise on the High Frontier

NATHAN C. GOLDMAN

BALLINGER PUBLISHING COMPANY
Cambridge, Massachusetts
A Subsidiary of Harper & Row, Publishers, Inc.

Copyright © 1985 by Ballinger Publishing Company. All rights reserved. No part of this publication may be reproduced, stored in a retrieval system, or transmitted in any form or by any means, electronic, mechanical, photocopy, recording or otherwise, without the prior written consent of the publisher.

International Standard Book Number: 0-88730-003-0

Library of Congress Catalog Card Number: 84-16761

Printed in the United States of America

Library of Congress Cataloging in Publication Data

Goldman, Nathan C.
  Space commerce.

  Bibliography: p.
  Includes index.
  1. Space industrialization—Government Policy—United States.
I. Title.
HD9711.75.U62G65   1984   338.4   84-16761
ISBN 0-88730-003-0

To my mother, Hilda Goldman, and to the memory of my father, Reuben Goldman, who gave me intelligence to question and courage to challenge.

Bibliography	179
Index	183
About the Author	187

CONTENTS

List of Figures and Tables   xi

Acknowledgments   xiii

PART I   *SPACE: THE COMMERCIAL FRONTIER*

    Chapter 1
    The Arena of Space   3

    Chapter 2
    The Space Powers   9

PART II   *SPACE COMMERCE*

    Chapter 3
    Space Commerce in Perspective   31

    Chapter 4
    Space Transportation   39

Chapter 5
Telecommunications 55

Chapter 6
Remote Sensing 75

Chapter 7
Manufacturing, Mining, and Energy 89

PART III   SPACE BUSINESS AND SPACE POLICY

Chapter 8
Domestic Space Business 103

Chapter 9
U.S. Policy toward Space Commerce 127

Appendixes

A — Treaty on Principles Governing the Activities of States in the Exploration and Use of Outer Space, Including the Moon and Other Celestial Bodies 145

B — Convention on International Liability for Damage Caused by Space Objects 149

C — Convention on Registration of Objects Launched into Outer Space 155

D — Agreement Governing the Activities of States on the Moon and Other Celestial Bodies 157

E — Convention for the Establishment of a European Space Agency 163

F — Section 305, NASA Act of 1958, as Amended. Property Rights in Inventions 169

G — NASA Patent and Data Policy for Shuttle Services Provided to non-U.S. Government Users, 14 CFR, Section 1214.104 171

H — H.R. 3942: A Bill to Provide for Commercialization of Expendable Launch Vehicles and Associated Services, House of Representatives 173

LIST OF FIGURES AND TABLES

Figures
3-1	Kennedy-Goldman Model of Space Commerce	33
6-1	EDC Sale of Landsat Imagery Frames	83

Tables
1-1	Policy Process	6
2-1	World Space Launches, 1957-1981	10
4-1	Vehicles and Payloads, 1983	42
4-2	Private Space Vehicles	45
5-1	Intelsat—National Shares	61
5-2	Inmarsat Member Nations and Their Percentage Investment Shares	62
5-3	International Satellites	64
5-4	U.S. Communications Satellites	66
6-1	The Remote Sensing Market—Public/Private Goods	78
6-2	Landsat Users, by Percentage	80
6-3	History of Remote Sensing Market Projections	82
6-4	Planned Sensors Compared	87

7-1	Twelve Typical Candidate Pharmaceutical Products for Space Manufacture	94
7-2	Space Manufacturing—Companies and Proposed Projects	96
8-1	Aerospace Industry Sales by Customer, Calendar Years 1967-1981	104
8-2	Shuttle Contractors	107
8-3	Selected Major Aerospace Corporations and Area of Expertise	112
8-4	1979 Statistics for Selected Aerospace Companies	114

ACKNOWLEDGMENTS

I would like to thank Dean George Kozmetsky, Director of the Institute for Constructive Capitalism, at the University of Texas at Austin, both for the concept of this work and the support to finish it. The book's format came out of a discussion with the Dean about the direction of my research. As the E.D. Walker Centennial Fellow for the Institute, I have received financial, technical, and intellectual support to complete this work. In addition to Dean Kozmetsky, I would like to thank and acknowledge the advice and help of Ray Smilor, the Associate Director of the Institute, and of Michael Gill and Sten Thore. I would like to give special credit to Marilyn Kister and Linda Teague, who typed the manuscript and to Rebecca Schorin, who read it for style. I would also like to thank Dr. Jack Kirwan, an economist at the University of Arizona, and Oliver Hennigan, Jr., a space analyst for U.S. Aviation Insurance Group, who read the manuscript with a friendly but critical eye. Of course, I accept all blame for remaining errors, although I offer in advance the rapid growth of space commerce as a partial excuse.

Nathan C. Goldman
Austin, Texas

I SPACE
The Commercial Frontier

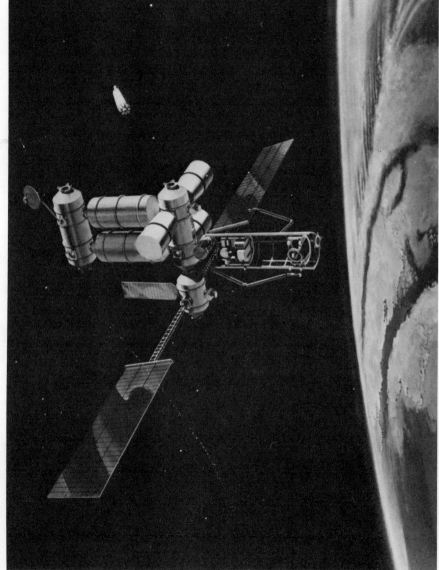

1 THE ARENA OF SPACE

In the last ten years space commerce has grown from scientific fantasy to economic reality. Most people have not yet awakened to this reality. This book is an alarm to wake the public to this second dawning of the space age.

It is a dawn full of challenge and change. The rapidly expanding literature on space war and space industrialization testifies to the growing importance of space for the nation and for the world. Adults who grew up during the era of Sputnik and Apollo (1957–1972) retain, however, the older cold war stereotype of space exploration as unproductive, expensive, and dangerous. This stereotype may have reflected reality in the 1960s, but by the 1980s, it had lost much of its veracity.

As early as the 1960s, the United States and Soviet space programs had begun to produce and develop direct and "spin-off" technologies that were soon integrated into the economies of the space powers. Moreover, space began to develop as a medium for commerce in its own right. Communications and remote sensing satellites began to rival and to surpass their air- and earthbound alternatives. As the 1980s began, space was also being touted because of near-zero gravity and other attributes, as a future site for the manufacture of materials that are expensive or only experimental on earth but with potentially large markets. These materials, such as pharmaceuticals and semiconductors, would be produced in orbiting, automated factories before this decade is out. Two other promising ideas for space commerce likewise have received substantial preliminary thought and experimentation:

- Solar energy transmitted to the earth to power homes and industries
- Celestial ores—from the moon or the asteroids—mined and refined in space for use initially in space and later on the earth

This is the new economic reality and potential of outer space. Other countries recognized this potential and made early efforts to enter the space race, but only in the 1970s did they achieve results worthy of note. Western Europe coordinated and consolidated most of its space efforts in the European Space Agency (ESA). Japan and China, at the same time that the United States landed its first men on the moon, launched their own first satellites into orbit. By the end of the 1970s ESA, Japan, The Peoples Republic of China, and India, in addition to the United States and USSR, had launched their own satellites. By the early 1980s, Brazil and other countries were also developing the scientific-industrial capacity to build and launch satellites. Moreover, almost every nation had been affected directly by the Space Age. More than 100 nations belonged to Intelsat, the International Telecommunications Satellite Consortium, and every nation in the world now used some weather, remote sensing, or communications satellites.

Meanwhile, the dark side—the military aspect of space—has continued to expand. Outer space, from the beginning, has been used for military communications and espionage. Over the past twenty-five years, however, these applications have been so refined that the U.S. military has become heavily dependent on these space applications. Instant communications between headquarters and field locations and instant and accurate spotting of friendly and enemy troops have given an important "force multiplier" to the U.S. armed forces, increasing dramatically the overall effectiveness of a few troops.

Despite our early hopes for peace in space, even expressed in treaties, this space-based assistance to our armed forces has strategically become an important target for our potential enemies. The Soviet Union developed antisatellite weapons as early as the late 1960s; and U.S. civilian and military satellites seem destined to be attacked in any conflagration of the major powers and need, therefore, to be defended from attack. By the early 1980s, the U.S. government was spending more money on the military in space than it spent on the civilian space programs. Much of these funds would be expended on hardening (shielding) military satellites.

The U.S. Navy and U.S. Air Force have both established Space Commands to direct space efforts. Much of their budgets are expended on passive "hardening" to protect the satellites from attack. Strategists in the U.S. War Colleges are developing defensive and offensive plans for space, and advanced plans for military space hardware include designs for a space plane, antisatellite missiles and ultimately space-based defensive (if not offensive) weapons aimed at air, sea, land, and space targets.

President Reagan has expressed great interest in space weaponry as an opportunity to replace the Mutual Assured Destruction (MAD) strategy with a system that will defend against nuclear weapons. This technology, according to some sources, would permit the United States to leapfrog the Soviet advantage in quantity of arms. Alternatively, this defensive capability would reduce the value of present offensive nuclear weapons and may encourage real arms reduction negotiations in the future.[1]

Other works deal adequately with the unfolding area of military use of space. This book considers instead space commerce, especially private enterprise in outer space. The social scientist's questions concerning the distribution of power—Who gets what? Who wins? Who loses?—are considered here from both the economic and political points of view.

Another organizing principle this inquiry relies on is the sub-discipline of public policy. Public policy may be conceptualized and studied in three progressive periods (1) recognition of a concern or the need for a policy and preliminary formation of a policy, (2) response, or enactment of a policy designed to address the need, (3) assessment of the policy's effectiveness. (See Table 1-1.)

This book will focus on this first stage of public policy—the formation of an issue. This perspective is necessary because the factual situation of space commerce and formulation of space policy are so new and changing so rapidly that the possibilities, promises, and challenges of outer space are constantly reshaped by the technological and political events.

These events are quickly history. And as the space age matures, its children grow up and assume power, having different perspectives on

Table 1-1. Policy Process.

Perception	
Definition	
Aggregation	Recognition/Formation
Organization	
Representation	
Formulation	
Legitimation	
Appropriation	Response/Enactment
Implementation	
Evaluation/Appraisal	Assessment
Resolution/Termination	

Source: The left-hand column is excerpted from Charles O. Jones, *An Introduction to the Study of Public Policy*, 2nd ed. (North Scituate, Mass.: Duxbury Press, 1977).

the world than their parents had. In the early years of the space age, advocates still had to convince a skeptical public of each new possibility in space. By the 1980s, a generation gap had emerged over the issue of space. The advocates of space commerce and other uses of space technology still have to leap the barrier of initial disbelief inherent with older members of the populace. The generation under forty—just now assuming political and economic power—have no such impediments. Of course, it still must be convinced of the economic viability of a particular program, but not of its technical possibility. It does not share the psychological inhibitions of its elders.

This book recounts many of the events that are shaping this new postindustrial world, or shall we say, postindustrial solar system. As such, it is a study of public policy formulation at the threshold of a new phase of human history. The following chapters lay out the origins and structure of this space commerce, the space programs and policies of the spacefaring powers, and finally some of the policy considerations that will confront the U.S. decisionmakers both in government and in private enterprise as we enter this newest realm of high tech competition.

NOTE TO CHAPTER 1

1. See Daniel O. Graham, *The High Frontier: A New National Strategy* (Washington, D.C.: Heritage Foundation, 1982), and *We Must Defend America: A New Strategy for National Survival* (Chicago: Regnery Gateway, 1983). General Graham develops the concept of space-based weapons as the best solution to MAD, the theory of Mutual Assured Destruction. Daniel Deudney, Global Security: The Geopolitics of Peace, World-Watch Papers, no. 54, 1983, presents a thoughtful, well-researched counterargument to the high frontier. The balanced account of space military is presented by Thomas Karas in the *New High Ground: Systems and Weapons of Space Age War* (New York: Simon and Schuster, 1983).

2 THE SPACE POWERS

At least since the second International Geophysical Year (IGY) in 1955, outer space has been an area of both international competition and cooperation. The vast expense and risk of most space projects ensure some tendency to share the costs. This is especially true in the space sciences, which are not viewed as directly military or commercial applications. Even amid the tensions of a new cold war, the United States and the Soviet Union continue to cooperate on biological experiments aboard the Soviet Cosmos satellites and Salyut space stations. The United States and Europe (as ESA) cooperate in the Spacelab program; individual European nations also work with NASA, such as United Kingdom and the Netherlands, on the Infrared Astronomy Satellite (IRAS). Japanese experiments in space manufacturing ride aboard the U.S. Space Shuttle. Moreover, every space power in the world has become involved in some way in the scientific exploration of Halley's Comet, which returns to the vicinity of earth in 1986.

The movement into space nonetheless has its extremely competitive attributes, which have continually expanded with the maturation of the technology. And the United States and the Soviet Union are not alone in this competition. Table 2-1 depicts both the deceleration of U.S. space launching and the acceleration of competitors in the rest of the world.

Table 2-1. World Space Launches, 1957–1981.[a]

Year	Launches						
	USSR	United States	Japan	China	Italy	ESA	India
1957	2						
1958	1	5					
1959	3	10					
1960	3	16					
1961	6	29					
1962	20	52					
1963	17	38					
1964	30	57					
1965	48	63					
1966	44	73					
1967	66	57			1		
1968	74	45					
1969	70	40					
1970	81	29	1	1	b		
1971	83	31	2	1	1[b]		
1972	74	31	1		b		
1973	86	23					
1974	81	22	1		2		
1975	89	28	2	3	b		
1976	99	26	1	2			
1977	98	24	2				
1978	88	32	3	1			
1979	87	16	2			1	
1980	89	13	2				1
1981	98	18	3	1		2	1
1982	102	18	1	1			1
Total	1538	796	21	10	4	3	3

a. Besides the countries and organizations listed that currently have space launching capability, three other countries have successfully orbited payloads in the past: Britain made one successful space launch of a single payload in 1971 and had one failure in 1970. France had its own launch vehicle (Diamant) from 1965 through 1975 during which 10 successful launches of twelve payloads were made; there were two launch failures (three payloads). Australia made one launch of one payload in 1967.

b. Italy launched in each of four years footnoted a payload for the United States that is also included in the U.S. count, making a total of eight Italian launches and a net of 792 U.S. launches. The table makes no attempt to show separately other international coopera-

Table 2-1. continued

	Payloads to Orbit						Payloads to Moon and Beyond	
USSR	United States	Japan	China	Italy	ESA	India	United States	USSR
2								
1	5							
9							1	3
3	16						1	
7	35							1
25	55						4	1
19	62							1
37	69						4	2
69	93						4	7
47	94						7	5
74	78			1			10	1
81	61						3	4
76	58						8[c]	
98	36	1	1	c			3	5
107	45	2	1	1[c]			8	7
106	33	1		c			8	3
123	23						3	7
111	27	1		2			1	3
136	30	2	3	c			4	4
139	33	1	2				1	2
125	27	2					2	
138	34	3	1				7	4
126	18	2			1			
132	16	2				1		
146	20	3	3		5	1		2
1928	977	20	11	4	6	2	79	66

tive payloads launched by the United States, the USSR, France, and Italy for 18 countries or international organizations; these launches and payloads are included in the data listed in the table.

c. Additionally, the United States sent one piece of debris to escape as part of an earth orbit mission.

Source: Table 1a, Charles S. Sheldon and Marcia S. Smith, *Space Activities of the United States, Soviet Union and Other Launching Countries/Organizations* (Washington, D.C.: Library of Congress, Congressional Research Service, 1982), update from Aerospace Facts and Figures 1983/84.

Although the Soviet Union remains the chief competition of the United States in space, other nations have begun to develop their own space potential in the last decade. Much as Spain and Portugal opened up the oceans and the seas in the fifteenth century and were soon accompanied in the New World by other seafaring nations, the United States and Soviet Union have been joined by several late-starting space efforts. Europe and Japan rank well behind the United States and Soviet Union but well ahead of India, China, and Brazil in this new space race.

This chapter will review the space program, policy, and potential of the other members of the Big Four in space—Europe, Japan, and the Soviet Union. The discussion will provide much of the context for the later chapters, which deal with the policy options that the United States faces in its competition to retain its position in the space market. But first let us examine a few of the motivations—some rational, some less so—that impel humanity on its seemingly instinctive and ineluctable drive into outer space.

MOTIVATIONS FOR SPACEFARING

Adventure

While this human emotion—a craving for adventure—may not drive a nation into space, it is a powerful force directing individuals and may be the ultimate, if unquantifiable, reason that humanity has ventured beyond the earthly crib. We seek to be more than we are. We wonder and wander. Wonder may be our most precious asset; it has often paid off for humanity in the past.

Military Power and Prestige

In the 1950s and 1960s military advantage and national prestige were the primary motives for the great space race. The Soviet Union and the United States pursued these related goals in the context of global competition, but smaller powers likewise may have military or prestige motives for venturing into space. The technology to orbit satellites is the technology to launch intermediate-range and long-range intercontinental ballistic missiles. When combined with a nuclear weapons capacity, such as China and possibly India and Brazil will have, the new space power has established itself as a regional if not a world power. For example, even though India has denied any intention to combine its nuclear and space technologies, Pakistan, long an adversary, very much fears this capability.

National pride *and* prestige *are* related, but separate, incentives for venturing into space. They have military, political, and economic ramifications. A Chinese official reflected this motive for space flight when he stated that the "People's Revolution" would not be left behind but would be in space with the superpowers. In 1970, the first Chinese satellite broadcast a song, "The East Is Red," and boasted of China's enhanced status in the community of nations.

Similarly, Brazil has long been touted as a coming superpower. Brazil's space effort has reflected its conscious pursuit of that image. This use of space technology as a means of acquiring world recognition has a direct parallel in the sixteenth and seventeenth centuries' age of exploration. Then, a nation desiring status as a world power might establish colonies in the new worlds. These colonies, however, usually proved a national burden. Rarely were these efforts financially beneficial; indeed, they were often difficult and expensive to maintain and defend. Nonetheless, they were an important status symbol. Spacefaring may likewise become an expensive symbol of our age. Although space applications are undeniably beneficial, most of those benefits can be acquired much more cheaply by purchasing satellites or launches from the established space powers. Yet, spacefaring, like seafaring, remains an expensive gamble taken by nations wishing to acquire greater prestige and status.

Economic Development

Many persons consider prestige and military prowess to be tainted justifications for space exploration, especially in light of the many poor and hungry of the earth. This taint seems especially strong with regard to the space programs of countries such as India and Brazil, where a large portion of the populations subsist in deprivation. A closer analysis of the implications of space applications for these people and nations, however, may dissipate the validity of this argument from space critics.

Communications and remote sensing satellites have already played a major role in the economic and national development of many Third World nations. Satellites bridge—or better, hurdle—the barriers of distant or difficult terrain. China, India, and Brazil are paradigms for the geographic problems that can be remedied by satellites. Not surprisingly, these applications are the major focus of all three space efforts.

In areas in which traditional earthbound communications technologies would be expensive if not impossible to link together, satellites can broadcast not only telephone but visual transmissions with ease. The applications seem endless.

Experiments in Canada and India have demonstrated that televised educational programming could be transmitted by satellites from the capital city to community receivers in the countryside. With the shortage of trained teachers, satellite instruction may be a relatively inexpensive way to teach an entire generation of children. Similarly, satellite communications can stretch the thin supply of doctors, linking nurse practitioners in the rural areas to the urban hospitals. Everything from diagnosis to operations can be performed by the nurse practitioners under the supervision of the centralized hospital staff.

Communications has often been juxtaposed as an alternative to transportation. If satellites can serve that function cheaply, the need for air, sea, or land transportation becomes less pressing. Satellites affect, ironically, both the communications and transportation sides of the equation. Remote sensing satellites have bolstered the transportation side of this equilibrium. Brazil, especially, has used remote sensing data to choose the best trans-Amazon route. These data have already permitted an accurate mapping of the entirety of the Amazon basin; it was discovered that the aerial maps had been incorrect by many miles.

Remote sensing has served Third World nations in other ways. Satellite meteorology has saved lives and property by early warnings of monsoons, hurricanes, and flooding. Farmers can even use remote sensing to identify crops in need of irrigation at an earlier time; moreover, national bureaus for agriculture can use the data to predict crop yields and generally to rationalize their farm programs. And sensing data can also be used to find mineral deposits, including oil or precious ores.

In a developing nation divided by languages, cultures, and geography, these satellite applications can help that nation leap several stages in the building of a national identity and infrastructure.

Science and Technology

Developed and developing countries have also considered going into space because the development of their own technological and industrial base allows them to compete better in the modern economic marketplace. Third World countries, constructing their own satellites and rockets, may find greater expense in the short run than in buying the hardware from the established space powers; it may even seem like reinventing the wheel. Yet, foresighted decisionmakers consider this expenditure a good long-term investment in national economic independence. The simple fact is that space exploration will generate the market for a host of domestic industries such as electronics, metallurgy, and computers which will permit the developing nations to overcome their technological subservience to other powers.

Commerce

Every advantage of space exploration discussed so far could be included in space commerce. The term *space commerce* is used here in a limited, special sense. Hereafter, space commerce will refer to the market for space goods and services including space transportation, communications, and remote sensing satellites; manufacturing in space; and eventually the transmission of solar energy back to earth by satellite and the mining of celestial bodies. Either through private enterprise or nationalized efforts, nations have begun to compete in earnest for the space markets valued in the tens of billions of dollars. In the following chapters, we will discuss this competition, most notably among the United States, Western Europe, Japan, and the Soviet Union.[1]

THE SPACEFARERS

Europe

As early as the 1950s, European countries such as the United Kingdom and France had developed small-scale space efforts. After World War II, however, Western Europe had faced up to the reality of the last half of the twentieth century: Only as a unit could Western Europe hope to compete with the superpowers militarily, politically, scientifically, and economically. The European Common Market and the European Parliament are outgrowths of this recognition, as is the development of a European Space Agency.

In the 1960s the nations of Western Europe created two projects: the European Launch Development Organization (ELDO), to build a European rocket; and the European Space Research Organization (ESRO), to conduct basic research in the space sciences. In 1975 these two organizations were merged into the European Space Agency (ESA). By 1983 there were eleven members and three associate members: Belgium, Denmark, France, Ireland, Italy, the Netherlands, Spain, Sweden, Switzerland, West Germany, and the United Kingdom, and associates Austria, Norway, and Canada.

Although ESA's goal was a "Europeanization" of local space efforts, national rivalries and interest created problems from the beginning. The German and French representatives were unable to agree on which nationality would have the honor of being represented in the person of ESA's first director. An Englishman, Roy Gibson, was chosen interim director and later became its first permanent director.

The ESA budgetary arrangement likewise recognizes the basic problem of trying to coordinate the policy of a dozen different nations. The ESA charter provided for a segmented budget; certain operating expenses were mandatory and fixed among the nations. The charter also permitted nations to propose specific programs, such as a communications satellite or launch vehicle. Members could voluntarily join the program as well as decide how much of the burden to bear. [Excerpts from the ESA Convention (1975) are presented in Appendix E.]

Transportation. In addition to the mandatory contribution, the ESA charter authorizes these voluntary contributions for optional programs. The most important and successful voluntary ESA program to date has been the Ariane family of rockets. Begun in 1975, Ariane was one of the original ESA projects. By far the major contributor to this program was France, with nearly 60 percent (West Germany contributed another 20 percent). The French space agencies and French companies therefore received 60 percent of contracts related to Ariane. And when ESA created Arianespace as a semiprivate corporation to market and operate Ariane, French banks and other institutions received 60 percent of the shares. Meanwhile, ESA and the French National Center for Space Studies (CNES) continued to conduct the research and improvements of the rocket. This history and division of labor resembles the history and the division of labor between NASA and the Communications Satellite Corporation, Comsat, after 1962.

Ariane competes directly with the U.S. launch vehicles. It operates a Washington, D.C. office and has booked passage for the satellites of several U.S. companies, including GTE, Western Union, and Southern Pacific Communications. By 1982 Arianespace had booked more than 30 orders for launchers—at over $20 million a flight.[2]

The Ariane 2 and 3 presently can put one 2,580 kilogram satellite or two, 1,195 kilogram satellites into geosynchronous orbit. By 1986 customers of Ariane 4 will have four options to place satellites ranging from 2,000 to 4,300 kilograms into geosynchronous orbit. With the recent construction of a second launching pad at its French Guinea site, Arianespace will be able to achieve a one-month turnaround and a dozen launches a year.

The head of the new company opened his leadership with a speech stating that he sought to capture one-quarter of the market, to launch at least 50 of the 200 satellites to be orbited in the coming decade. Arianespace has pursued this target with marketing and financial prowess. German and French bankers require only 20 percent down and payments at 12.5 percent interest beginning six months after the flight. Presently U.S. pricing requires the launch to be paid off before take-off.[3] Ariane, likewise, receives its money up-front, but requires 20 percent of this funding from the customers. Ariane helps customers obtain these loans; the U.S. government, conversely, offers no similar assistance to shuttle users, who must appropriate their own financing. Fortunately for the United States, world interest rates by late 1983 were more favorable for the United States, so the marketing success of Arianespace had been partially defused.

Communications Satellites. European nations, individually and in consortia, have made much progress in space applications. Europe has benefited directly with improved telephone and television. Although Europe is not benefited as directly by direct broadcast satellite (DBS) technology, some countries in the Third World have expressed an interest in better communication systems. Europe would be able to enter this new market (at more than $20 million a contract) in competition with U.S. and Japanese satellite companies.

ESA's Orbital Test Satellite (OTS-1) failed in its first effort in 1977, but OTS-2 succeeded in 1978. This enabled ESA to experiment with a 14/11 gigahertz (GHz) satellite, so named because the satellite receives its broadcast frequency at 14 GHz and returns it to earth stations at 11 GHz (frequency popularly referred to as the Ku-band). The Interim Eutelsat Organization used the OTS-2 for several years. Eutelsat became more fully operational with the launch of the European Communications Satellite (ECS-1) on Ariane in June 1983. ECS-1 has twelve 14/11 GHz transponders for private and business communications. ECS-2 and ECS-3, with launchings in May 1984 and August 1985, will complete the system.

Eutelsat was created by the seventeen members of the European Conference of Postal and Telecommunications Administration (CEPT) in 1977. By May 1983, 18 states had ratified the convention establishing the organization. The Federal Republic of Germany joined in October 1983 and assumed almost 11 percent of the finances. Spain, Greece, and Yugoslavia were parties to the interim agreement but had not yet joined the definitive agreement.[4]

A more recent voluntary program for ESA has been the Olympus Satellite for European Television. Olympus will be operational in 1986, experimenting with (DBS) broadcasting directly to European viewer's homes. France and Germany, however, did not join this project because of their own endeavors. Instead, these two countries have operated separately from ESA to develop the Symphonie satellite in the 1970s and more recently the TDF-1 (a DBS) and TV-Sat in the 1980s to serve their telecommunications needs. National and private companies (such as the Luxembourg Company of Telediffusion) have also begun to compete for satellite telecommunication.[5]

Remote Sensing. European nations have begun to compete with the U.S. Landsat and other sensing satellites in remote sensing. The recent Eumetsat Convention provided the organization and the planning for a three Meteosat network for weather forecasting operating by 1987. The French CNES has already developed the SPOT (earth observation satellite) to be launched in 1985.[6] The French government will cede the worldwide marketing of its remote sensing data to a new corporation, Spot Image. The corporation will have CNES and other French institutions as its primary stockholders; Belgium, Sweden, and international banks will control the remainder. Spot Image plans to make its ground stations compatible with both Landsat and SPOT satellites so that its customers will receive the best of both systems.[7]

In 1983, the shuttle flew the German Shuttle Pallet Satellite (SPAS-01) to test out a civil remote sensing device called the *pushbroom scanner*, or Modular Optoelectronic Multiple Scanner (MOMS). The German company Messerschmitt-Boelkow-Blohm (MBB) built it. In late 1983 MBB joined with the U.S. Comsat and the Stenbeck Group, a Swedish investment group, to form the Sparx Corporation. Sparx will compete with Landsat, Spot Image, and Space America in the remote sensing markets. It will include a fleet of about ten SPASs, and the program will cost less than $100 million. Comsat later dropped out of the program and its other remote sensing interests due to an internal consolidation.[8]

Manufacturing. The major European effort in space manufacturing revolves about the future of Spacelab. One of the early voluntary ESA programs, Spacelab is the space laboratory that flies in the U.S. Space Shuttle and can support two astronauts for a week. West Germany was the largest contributor to this program and has dedicated an entire Shuttle-Spacelab mission in the middle 1980s to science and manufacturing in space.

MBB's SPAS-01 may also have potential to be the forerunner of an automated, retrievable space factory. The craft, produced with no government assistance, is reusable and will be leased to private parties for remote sensing or materials processing. It can be tethered or a free flyer and was successfully tested on the Shuttle in 1983.

These two options have begun to emerge as competing methods of manufacturing in space—the automated free flyer versus the staffed space station. The cheaper cost of non-human-tended factories will be weighed against the need for human intervention to determine which products will be produced by which mechanism. ESA has begun to develop such a retrievable satellite, called the Eureca, as a new optional program in 1984.[9]

European Policy. The ESA charter provides that an important goal of its programs will be to

> Improve the worldwide competitiveness of European industry by maintaining and developing space technology and by encouraging the rationalisation and development of an industrial structure appropriate to market requirements making use in the first place of the existing industrial potential of all member states...[10]

ESA has indeed fulfilled this mandate as regards European industry. Aerospatiale, a French company, has been active in both the construction of Ariane and of satellites. Another French company, Marta, has developed an expertise in remote sensing and communications satellites. Italy has followed a policy of being a secondary but substantial contributor to most ESA voluntary projects; this procedure assures Italy of substantial contracts for its industry, especially in the electronics for spacecraft.

Because the prime contractor for a new project is generally quite large and multifaceted, consortiums or groups of companies often pool expertise and resources and put in joint bids for the job. For instance, an industrial group composed of Marta, Aeritalia (Italian), British Aerospace, ERNO (West German), Fokker (Dutch), INTA (Spanish), and SAAB (Swedish) won the ESA contract for a series of astronomy satellites against a consortium headed by Aerospatiale (French).

The future of European space efforts depends on ESA experiences with Spacelab and other space applications as well as on the centrifugal tendency for European countries to proceed alone or in small groups. The French and Germans, for instance, developed their own communications satellites and are not members of Eutelsat. The European countries are attempting to position themselves to compete for the world markets in space applications—goods and services—as they become economic as well as technical realities. Space manufacturing is just the most farsighted of these efforts. But Europe still is struggling with whether to compete as a unity or as individual nation states.

Japan

In the context of the 1955-1957 International Geophysical Year (IGY), Japan's Institute of Industrial Science (IIS) at the University of Tokyo began to develop a series of sounding rockets. The Japanese prime minister established a National Space Activities Council in 1960 to advise him on these nascent space matters. During the first half of the decade, the University of Tokyo constructed the Kagoshima Space Center; also, the University's IIS and the National Aerospace Laboratory merged into the Institute of Space and Aeronautical Science (ISAS). Later, the ISAS became the Institute of Space and Astronautical Science and was transferred to the government's Ministry of Education; it continued to conduct research and launch science satellites from Kagoshima.[11]

Meanwhile, in 1957 a Science and Technology Agency had been set up inside the government. The Space Development Office of the Agency became the National Space Development Center. By 1966 it had begun to develop the Tanegashima launch site for larger rockets.

1968-69, the years of the U.S. lunar triumph, were the years of the fundamental reorganization of the Japanese space program from a research institution to an operational program. The National Space Activities Council ended its eight-year existence and was replaced with the prime minister's Space Activities Commission. The Commission published its first Space Development Program in 1969. Also in 1969 the National Space Development Center was dissolved and replaced by the National Space Development Agency (NASDA).

Although heavy space launches were NASDA's domain, the first satellite launched by Japan was the ISAS's Osumi. ISAS followed with other science payloads throughout the decade.

Today NASDA uses its launch site at Tanegashima for space applications and larger science missions. Hideo Shima was the father and first president of NASDA. In 1977 NASDA made Japan only the third country in the world to achieve a geostationary orbit. The Kiku-II was an experimental satellite but was the harbinger of others to come.[12]

Although Japan has been developing its domestic aerospace industry since the 1950s, until the 1980s the Japanese rockets—the N-series, were licensed designs based on the McDonnell Douglas Delta.

Japan therefore could not compete for worldwide launching business because of restrictions in its licensing agreements. Japanese authorities blame American workmanship for several launch failures in the late 1970s. For these two reasons Japan has been considering a basically Japanese-designed vehicle that would be legally able to enter the competition for launch services.

Even so, the H-1 rocket of the late 1980s will still only have the capacity to launch 1,120 pound payload into geostationary orbit (an improvement over the N-2's 800 pound limit). A policy dispute has resulted in Japan over whether NASDA will launch its projected heavy satellites in the U.S. Space Shuttle in the early 1990s or will delay the building of large satellites until Japan acquires the means to launch them near the year 2000 with an updated H rocket capable of 4,000 pounds payload. Such dilemmas are endemic to any policy of autarky, but since there is also a market in selling large satellites to other nations (India's *Insat*, Indonesia's *Palapa* are examples), it seems likely that Japan will build the large satellites and launch them, for a while, from Shuttle.[13]

Space Applications. Japan has also undertaken an extensive program of near-earth applications. Japan has major plans for remote sensing, including weather, environment, and maritime satellites. In 1981 the Japanese launched the Geostationary Meteorological Satellite (GMS), which took part in a world weather study together with U.S. and ESA satellites. In 1986 Japan plans to launch its Earth Resource Satellite (ERS-1). By 1990 Japan will launch its J-ERS-1 satellite, with new radar and microwave sensor technologies. A planned Earth Resource Observation Analysis Center will phase out Japanese reliance on U.S. Landsat information.

The 1985 Maritime Observation Satellite (MOS) will boast a new scanning device, which is mechanically superior to Landsat's multispectral scanner. It will have nearly 100 percent scanning efficiency, high speed, and will not require radiation cooling. MOS will assume a sun-synchronous, low earth orbit to study oceanic conditions—an important application for Japan's fishing and shipping interests.[14]

Communications Satellites. Japan has actually taken the lead in some advanced technologies in communications satellites.[15] The Communications Satellite (CS-2b) was launched from Tanagashima in 1983. In addition to its C-band 6/4 GHz transponders, CS-2b also has 6 Ka-band, 30/20 GHz transponders. Japan thus became the first nation to have operational 30/20 channels. With crowded 6/4 GHz and anticipated crowding in Ku-band 14/12, this technology prepares Japan to be able to develop techniques to overcome problems with 30/20 transmission and then to market these high-frequency satellites worldwide.[16] In the middle and late 1980s, Japan also plans a series of broadcast satellites (BS) that will exploit these new techniques to overcome the chronically poor TV reception on the Japanese archipelago.[17]

The Japanese government and private industry cooperate closely in communications. The Telecommunications Satellite Corporation Telesat-Japan is the private company that represents the nation in Intelsat and Inmarsat. Supervised by the Ministry of Posts and Telecommunications, it serves in a capacity comparable to Comsat in the United States.[18]

Materials Processing and Manufacturing. NASDA is interested in the commercial potential of the Shuttle and Spacelab. The first Spacelab flight includes a Japanese experiment. Japan hopes to study manufacturing of alloys, semiconductors, and medicine on future flights. In preparation for these studies, Japan has tested materials processing from a series of sounding rockets.[19] These TT–500A tests have produced nickel alloy and semiconductors during the seven minutes of microgravity. Japan has reserved one-half of a U.S. Space Shuttle payload to continue these experiments.[20]

Looking to the future, Japanese officials have expressed a desire to participate with the United States and Europe in a manned space station that may be able to conduct ongoing space manufacturing. Japan also is looking to its own manned reusable shuttle in the 1990s. This proposed craft would be orbited by a modified H–1 rocket; the crew of four could remain in a 240-mile-high orbit for two days.[21]

Japanese Industrial Policy. Japan has continued its policy to developing domestic high technology both for export and for self-sufficiency. The Japanese Ministry of International Trade and Industry (MITI) maintains a space group to coordinate industrial policy with the space policies devised by prime minister's Space Activities Commission.

Increasingly, Japanese automobile and heavy industries have moved into the aerospace market. By cooperating with foreign companies, the three major Japanese satellite manufacturers, Nippon Electric Corporation (NEC), Mitsubishi Electric, and Toshiba, have been able to acquire expertise and technologies to develop their own capacities in space. Presently NEC cooperates with RCA and Hughes Aircraft, and Toshiba is aligned with General Electric. The most widespread alliances involve Mitsubishi Electric, which cooperates with the American Ford Aerospace, the French Aerospatiale, and the West German Messerschmitt-Bolkow-Blohm (MBB).

Although the Japanese industries compete with each other in some respects, they cooperate with one another on research and occasionally on market strategy. In other high-technology fields, for example, the MITI's Agency of Industrial Science and Technology has been the primary mover in developing Japan's new "supercomputer."[22] Government and business have cooperated in the development of semiconductors. Also, five Japanese companies have created the Biotechnology Council on Research and Development to develop new technologies and markets in this growing area of knowledge.

Governmentally, Japan's Space Activities Commission produces and updates periodically its Outline of Japan's Space Development Policy. The Commission reports directly to the prime minister and thus allows the space bureaucracy to speak with one well-amplified voice directly to the top. In contrast, since the demise of the U.S. National Space Advisory Council, (NASC), in the mid-1970s, none can speak for the combined U.S. space effort at any level. Japan's coordination is an immeasurable but clear advantage of its space policymaking over its U.S. counterpart.

It must be noted that Japan has special historical reasons for its high technology and space orientation.[23] Japan has to import nearly 99 percent of its raw materials, including oil. Many of the wars that Japan has fought in the past century have had their origin in matters related to economic resources. Neil Davis has argued that security for Japan is not military (partially because of the U.S. nuclear umbrella) but economic in nature.[24]

Davis, however, finds that—just as for the other industrialized powers—space has yet to achieve the recognition of its potential by the Japanese public or the planners. Thus, while space commerce may have more push in Japan than in other nations, it does not receive (perhaps it does not yet warrant) the same attention as computers and other established markets.

The Soviet Union

The Soviet Union is viewed only as our competitor in ideology and the military; commercially, the United States and the Soviet Union have had little in the way of natural competition. Space commerce brings the two political rivals into direct economic rivalry for the first time. This aspect of space conflict, however, did not become apparent until the 1980s.

The Soviet Council of Ministers makes budget and policy decisions on space and defense in the Soviet Union. The Council supervises and relies on information from the Committee for Science and Technology and from the Academy of Sciences for its decisions. The Academy, in turn, supervises the Commission for Exploration and Use of Outer Space and the Institute of Space Research (ISR). These two organizations actually plan and coordinate the space program. Unlike NASA's space missions in the United States, the Soviet ISR space missions are launched by the military. This Strategic Rocket Force has a quarter million members and trains and supervises the cosmonaut corps.[25]

Although the Soviet space effort has traditionally been more militarily oriented than the United States', the Soviet Union has devoted much effort to the science and near-earth applications of space. After Apollo, the U.S. space effort flagged, but the Soviet program continued to a slow but consistent buildup. The Soviet Union, by the 1980s, was investing more than 2 percent of its gross national product on space; that is twice the U.S. investment during Apollo. The Soviet Union outspends the United States in space at an estimated rate of $20 billion a year compared to the United States' $14 billion a year.

The major manifestations of this expenditure have been nearly 100 Cosmos satellite launches a year along with the Salyut series of space stations, which Soviet cosmonauts have occupied on a semipermanent basis since the late 1970s. The Soviet Union has conducted years of research on materials processing, including metallurgy in space.

The Soviets have developed three other spacecraft that supplement the Salyut and might facilitate its commercial potential in the near future. First, the Soyuz-T is the vehicle that will ferry cosmonauts to the Salyut, at least until the Soviet space plane (a shuttle) becomes operational in the late 1980s. The Soyuz design is almost 20 years old; although not sophisticated, it still works.

In the late 1970s, the Soviets began to use their Progress Rocket, a robotic transport that refueled and resupplied the station. Since 1981 the Soviet Union has been able to dock Cosmos satellites with the Salyut. These satellites give the Salyut another module for living quarters or for science and manufacturing facilities. Most important, the craft can ferry 2.5 times as many supplies as the Progress, and it can return more than 1,000 pounds to earth. This thousand pounds, obviously, could include valuable pharmaceuticals, semiconductors, or special alloys manufactured in space.[26] The near-vacuum, low gravity space environment might well facilitate the commercial production of such materials that are presently experimental or virtually unfeasible on earth.

Communications and Remote Sensing. The Soviets developed remote sensing later than the United States. Only in 1966 did the Soviets begin to receive frequent satellite pictures for weather forecasting. 1977 marked the first Soviet weather satellite in a sun-synchronous orbit, such as used in the U.S. programs. Since the late 1970s, the Soviets have launched dozens of small sensing satellites, but it is not clear which were military or civilian remote sensing. The Priorda Center stores the Soviet earth resources data much as does the U.S. Geological Survey's installation at Sioux Falls, South Dakota.

For their communications satellites, the Soviets have tended to forego geosynchronous orbit. Their Molniya is put into an orbit 63° to the equator so that its low point is over the southern hemisphere. This elliptical orbit puts the satellite over the Soviet Union for a much longer period. Eight of these satellites permit the Soviets to have continuous coverage for communications. More recently, the Soviets have moved to the equatorial orbits with its Raduga (Ekran) and Gorizont series.

The Soviet Ekran satellite provides high-powered television and other telecommunications services. The Soviet Union also spearheaded the creation of Intersputnik, a consortium of Soviet Bloc and some Third World nations, that provides space-based telecommunications among the nations. This organization rivals and partially duplicates the Intelsat organization.[27]

Soviet Policy. The Soviet space muscle has clear military and science applications that have clouded the commercial potential of their space establishment. The Soviet Union has shown no propensity to commercialize its remote sensing or even communications satellites on the world market. Nor has the Soviet Union shown an inclination in this direction with its materials processing. But in 1983 the Soviet Union proposed to bid for the contract to launch the new Inmarsat satellites. The bid of $25 million dollars, is competitive with the Ariane, the U.S. Space Shuttle, and Titan alternatives.

It is unclear, however, whether this offer is a propaganda ploy or the first in a new economic competition for Third World and other international satellites. It should be noted in passing that the Peoples Republic of China made a similar offer in 1982 at the Unispace Conference to launch Third World payloads.

The logic of the Soviet Union's entry into world space commerce is strong. Trade is propaganda; it is also hard currency. The only surprise is that the Soviets had not entered this market earlier.

Conclusions

Analysis of the three other major world powers in space clearly demonstrates the importance of space commerce. The competition is building, and the rewards are many, both political and economic. The United States, both national and corporate, must formulate its plans in the context of this growing worldwide market and competition.

NOTES TO CHAPTER 2

1. *Space Flight*, British Interplanetary Society, 1977; "Convention for the Establishment of a European Space Agency," May 30, 1975; *Space Law: Selected Basic Documents*, 2nd ed., U.S. Senate Committee on Commerce, Science and Transportation, 95th Cong., 2nd sess., 1978.
2. ESA Brochures; *Aviation Week and Space Technology*, January 5, 1981, January 12, 1981, March 7, 1983, March 14, 1983.
3. *Satellite Communications*, September 1983; *Space Calendar*, July 4, 1983, August 15, 1983, September 19, 1983.
4. *Space Calendar*, October 17, 1983; *Aviation Week and Space Technology*, August 31, 1983.
5. *Civilian Space Policy and Applications*, Office of Technology Assessment, Congress, June 1982.

6. "Joint National Paper (of European nations)," 2nd U.N. Conference on Peaceful Uses of Outer Space, August 17, 1981; *Civilian Space Policy and Applications*, pp. 187-213.
7. *Aviation Week and Space Technology*, January 11, 1982.
8. *Aviation Week and Space Technology*, November 7, 1983.
9. *Space World*, October 1983; Space Calendar, December 5, 1983.
10. ESA Convention, note 1, p. 338, Article VII (1) (b).
11. Neil Davis, *Japan Times*; An Analysis of the Japanese Space Program: A Secure Foundation and a Need for Vision, unpublished manuscript, 1981, p. 42.
12. *Aviation Week and Space Technology*, June 23, 1981, July 27, 1981, August 10, 1981; *L-5 News*, September 1981.
13. *L-5 News*, September 1981; *Aviation Week and Space Technology*, April 6, 1981, July 27, 1981, August 10, 1981, August 24, 1981; *Space Calendar*, August 8, 1983.
14. *Aviation Week and Space Technology*, April 6, 1981, August 17, 1981, August 31, 1981, October 31, 1983; Davis, "Japanese Space Program," pp. 58-60.
15. *Satellite Communications*, July 1981, November 1981; *Space Age Review*, Autumn 1981.
16. Neil Davis, *Oriental Economist*, February 1983; Neil Davis, *Spaceflight*, 1983; *Satellite Communications*, August 1982.
17. *Space Calendar*, June 27, 1983, August 29, 1983.
18. *Civilian Space Policy and Applications*, p. 198; Neil Davis, *Japan Times*, August 8, 1982; Neil Davis, *Oriental Economist*, August 1982.
19. *Aviation Week and Space Technology*, August 24, 1981; *L-5 News*, September 1981.
20. Davis, "Japanese Space Program," p. 60.
21. *Aviation Week and Space Technology*, June 23, 1981, March 14, 1983.
22. Funds for a U.S. competitor has been deleted from NASA budgets.
23. Davis, "Japanese Space Program," pp. 1-10.
24. *Ibid.*, p. 10.
25. N. James Peter, *Soviet Conquest from Space* (New Rochelle, N.Y.: Arlington House, 1974); *Soviet Space Programs: 1976-1980*, Committee on Commerce, Science and Transportation, U.S. Senate, 97th Cong., 2nd sess., December 1982.
26. *Aviation Week and Space Technology*, April 6, 1981, June 2, 1981, June 14, 1982, July 19, 1982, August 2, 1982; *Space Calendar*, July 18, 1983; James Oberg, *Red Star in Orbit* (New York: Random House, 1981).
27. *Aviation Week and Space Technology*, August 10, 1981; Charles S. Sheldon and Marcia S. Smith, *Space Activities of the United States, Soviet Union and Other Launching Countries/Organizations* (Washington, D.C.: Library of Congress, Congressional Research Service, 1982), pp. 63-70.

SPACE COMMERCE

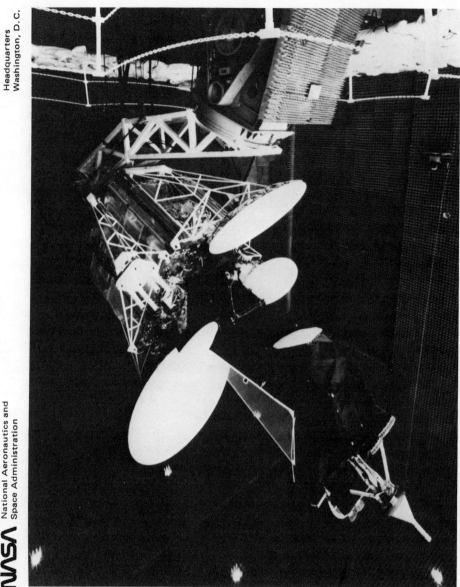

3 SPACE COMMERCE IN PERSPECTIVE

As early as 1958, the United States pursued the goal of space commerce, in which it would play a joint role with public and private enterprise. The act that created NASA reflects these goals: "The preservation of the role of the United States as a leader in aeronautical and space science and technology and in the application thereof to the conduct of peaceful activities within and outside the atmosphere."[1] This recognition of the potential of space is truly foresighted given the international crisis that prompted the acceleration of U.S. space efforts in the 1950s.

HIGH TECHNOLOGY

Since World War II, the industrialized world has experienced a series of new technologies that have revolutionized both the workplace and the structure of the economy. In the 1950s, the watchword was automation—as many unskilled jobs were lost to the machine. The 1960s followed with the computer as the scourge of the employed. Both automation and computerization, however, created millions of new jobs, so the major effect was to restructure rather than eliminate much of the work force. The scare for the 1980s may prove to be the robot revolution and perhaps in the 1990s, voice-accessed computers.

These new jobs, heavily high tech and mostly in the service sector, appear to be harbingers of a new economic system that may render much of the rationale of both capitalism and communism obsolete. Already a majority of the American "work" force are service, not production people. As we move into the era of self-replicating robots in the twenty-first century, the world will have entered a new era. If land and labor characterized the feudal period, and capital developed as an independent factor that defined the industrial age, technology seems to be a new fourth factor emerging in its own right, possibly even supplanting labor in the area of production!

Daniel Bell and David Apter have developed these ideas in the conception of the postindustrial age. But the ideas are not new. Karl Marx dealt with them in his discussions of alienation. For Marx, more workers lose their jobs and independence to the capitalist's reliance on the more efficient machine. Marx, however, could obviously not predict the resiliency and versatility of the economic and political structure to evolve new forms to fill the needs of the workers and of the society.[2]

The incentive mechanism in capitalism is both a positive and an efficient device, but personal freedom should rarely be sacrificed in the name of economy or society. This book deals with one aspect of the new technological society. Space commerce is already producing billions of dollars in revenue; most of its jobs will be earthbound, but first one and two, and then tens of workers will begin working and then living in space. Because of the unique attributes of space for commercial applications, this may well happen by the end of this century—that is, by the time the children now entering the first grade are graduating from college. It is not inconceivable that thousands of those children will spend their working careers in orbit about the earth. But it is the near-term, near-earth development of space commerce and related policy considerations that must be dealt with first.

SPACE COMMERCE: A MODEL

In the 1980s, space commerce is emerging as an important segment of the U.S. and world economy. Two seemingly contradictory trends—the integration of the space sector into domestic economies and the segmentation of space markets—suggest a strong future for business in space. Integration of the space segments—transportation, applications, manufacturing—means that all of the pieces are in place to be joined into a legitimate industrial sector. On the other hand, market segmentation connotes the continuing maturation and specialization of space hardware (goods) and services. The companies pursuing a specific space application can shop for the most appropriate launch and support services available, and the rocket launchers and other service suppliers can build and market their product to meet these specific demands.

An Integrated Model

Elsewhere, Michael Kennedy and I have begun to consider the implication of this integrated model of space commerce.[3] Reproduced in Figure 3-1 is the basic concept of the model. The diagram depicts at least two important facts about this new space economy.

1. Space transportation remains the most crucial variable, both in terms of costs and technological feasibility. In the noncommunist world, launch services (U.S., Japanese, European) can be valued at between $1 billion and $3 billion per year. The average year in the early 1980s includes maybe forty launches and sixty satellites; over half of these launches are American and most of the rest are European and Japanese.

2. The second point that can be drawn from this diagram is the interactive nature of the space economy. If the cost of launches is reduced, for example, the cost of the space applications

Figure 3-1. Kennedy–Goldman Model of Space Commerce.

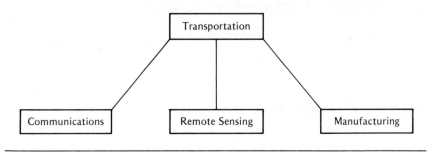

(other segments) will also go down. Obviously, the reduction in cost then makes the space-based industries more competitive with its earth-based alternatives. This logic should dictate a larger space segment, which in turn should precipitate more and cheaper launch facilities. Hence, the space economy requires an integrated and interactive model to understand space commerce in total and in its parts.

The most profitable sector of space commerce to date has been communications. Worldwide, this sector can be valued in the tens of billions of dollars. Every big name in telecommunications is intimately involved in satellite communications. AT&T, RCA, and the major broadcasting networks are the major participants in the international and domestic U.S. satellite industry. The industry itself has spawned a new set of big names all its own—MCI, Home Box Office (HBO), Cable News Network (CNN), Music Television (MTV)—all owing their present visability to these large structures in space. MCI succeeded initially through its microwave network, but has since purchased access to satellites.

Remote sensing satellites use similar technologies and deal with many of the same political and legal problems as communication satellites. As an economic market, however, remote sensing is still undeveloped. For the foreseeable future, the cost of operating the U.S. (NOAA operated) Landsat will exceed the revenues generated. Even the best estimate does not have a break-even until nearly 1990, at about $200 million each. The Congress has therefore been reticent about the Reagan administration's efforts to transfer these satellites (and the weather satellites) to the private sector in the near future.

The problem with this sector may be that remote sensing is both a public and a private good. The public good includes weather forecasting and general crop or geographic surveys. Such uses benefit all and are generally bought or at least subsidized by the state. The market value of this product is hard to determine by supply and demand considerations; moreover, the U.S. federal and state governments are and would remain its dominant consumer.

Remote sensing data also has a private goods aspect. Here, specific consumers of remote sensing data include the farmers who want to know about irrigation problems, weather, diseased crops, and the projected agricultural market. This example is closer to a public good. More purely a private good is the use that oil companies or fisheries will make of remote sensing data. The fisheries need specific data about weather conditions and even about location and size of schools of fish. Oil and mining companies need the data to decide on which geological sites to begin drilling or mining for resources or to bid for in government land auctions.

With better remote sensing data, these companies make a more accurate assessment of their costs and expected profits. The policy question remains, however, whether Landsat can operate as a private business that can meet the needs of the private consumers. Landsat was developed to prove a concept (remote sensing) not to compete in business. Structurally, it is at a disadvantage against its new competitors. The French government has already established a competitor, Spot Image Corporation, which will cater to the needs of both industry and nations.

Domestically, Space America—a consortium of Space Services, Inc., of America (SSIA, a Houston-based company) and American Science and Technology—are planning a series of remote sensing satellites that will be directed at the private goods aspect of remote sensing. Comsat and several other firms have expressed interest in all of Landsat or some aspect of the market. Comsat joined with the German company, MBB, in a new endeavor, Sparx Corporation, which will operate remote sensing payloads from the U.S. Space Shuttle. This new interest in and potential segmentation of the market suggest that remote sensing industries are in search of a market. The future of Landsat may be cloudy, but remote sensing looks very promising. It should be noted, however, that Comsat withdrew from all remote sensing endeavors in 1984 because of internal changes as well as the combative political climate.

The last sector of space commerce—manufacturing—is still largely a collection of reports on marketing and technology. These reports, especially the eight studies by aerospace companies in 1983, look extremely promising. They were conducted by NASA in order to identify the needs and uses of a proposed space station. By identifying products that can be manufactured in space, these companies found potentially billions of dollars in profit for goods and services provided by the stations. The promising products include pharmaceuticals, semiconductors, and alloys; services include construction of large structures in space and repair of satellites and other space facilities.

Now, the space commerce model completes its circuit: The entrepreneur in space, like any entrepreneur, must assess the potential of any venture. Their concerns are the cost of the project, the risk of the project, and the time lag and rate of return on the project. Space commerce, especially the speculative space manufacturing, have very high development costs, high risk, and long (if determinable) lead times and uncertain rates of return.

Cheap, reliable launch vehicles at least reduce the costs and risks of going into space, and less directly help with the other two considerations. Likewise, as more demand to go into space develops, the demand for (and supply of) launch vehicles should increase. Moreover, the specialized needs of space industries should act on the transportation (and space facilities) market to produce a proliferation of hardware to meet the needs of the industry.

Market Segmentation

The maturation of space hardware (goods) and services opens up even more options for space entrepreneurs. It is a chicken and egg situation: Which came first, the multitude of launchers or the proliferation of satellite applications?

Today, specialized rockets can respond to special applications. The multistage family of rockets permit the launch of different sized satellites to different orbits. For instance, SSIA is developing the Conestoga, which can put small, fully self-sufficient payloads into low earth orbit. Multiple stages and different sized boosters, such as the U.S. Delta, Titan, and Atlas-Centaur, permit the mating of satellites aimed for a specific purpose. Europe's Ariane and its family of boosters (discussed in Chapter 6) also fit into this category.

The U.S. Space Shuttle, however, creates so many options for users that it is a different kind of space transportation. Initially, the Shuttle can carry 65,000 pounds to low earth orbit as well as lift the largest satellites into a transfer orbit to geosynchronous orbit. The satellite owner in the late 1980s will have its choice of upper stage boosters for this trip, including the Centaur, Interim Upper Stage (IUS), Payload Assist Module (PAM), and private boosters discussed in a later chapter. At the other extreme, the Shuttle can carry tiny (200-pound) payloads as an afterthought, which permits even more versatile space activities. The availability of a human hand (and mind) on board to adjust instruments and to repair machines opens up new intricate procedures in space. Scientific experiments and manufacturing processes can be conducted in space. The reusability of the Shuttle also permits the placing of facilities in space and the retrieval of the facility or products from it.

Space Applications

As our learning curve in space improves, we will build more and varied space structures to carry out the growing list of tasks. Low earth orbit will be populated by remote sensing satellites. Stations also will occupy this orbit. Some will conduct science; others, military activities. Space factories, automated or staffed, will be designed to manufacture their product most efficiently. Another important use of low earth orbit space operations will be as a staging ground for the assembly and transportation of large space structures to geosynchronous orbit (for communications, energy, or observation) and as a launching pad for lunar and interplanetary probes. The infrastructure will then be established for a space-based industrial sector that will become fully integrated into the industry, society, and politics of the earth.

NOTES TO CHAPTER 3

1. *National Aeronautics and Space Act of 1958*, Sec. 102 (c) (5).
2. Daniel Bell, *The Coming of Post-Industrial Society: A Venture in Social Forecasting* (New York: Basic Books, 1976); David E. Apter, *The Politics of Modernization* (Chicago: University of Chicago Press, 1965).
3. "Politics and Economics of Space Commerce," in Paul Anaejionu, Nathan Goldman, and Philip Meeks, eds., *Space and Society: Choices and Challenges* (San Diego, Calif.: American Astronautical Society, 1984).

4 SPACE TRANSPORTATION

Getting into space is not only literally the first step in space commerce, but also the natural sector of the space market to discuss first. The dream of space travel is as old as the Greek, Semetic, and Indian legends describing flights to the planets and stars. The rocket could be found in ancient China. But the combination of the dream of space and the technology of the rocket was slow in coming. Yet when the combination occurred, rocket technology made huge advances.

Current space transportation is of several types. The United States, Europe, Japan, and the Soviet Union have space transportation systems that can compete for domestic and international cargo, such as satellites and passengers.

Space transportation may also be discussed in the context of the U.S. Space Shuttle and the expendables—the throwaway one-shot rockets, known as the ELVs, Expendable Launch Vehicles. Additionally, transportation may be discussed in the context of the sizes and types of payloads that are launched.

MARKET SEGMENTATION

The array of rockets and services and the increasing variation in satellite and applications has transformed the old one-stop shopping for the satellite or the rocket. Now satellite owners and users can shop for the service that best fits their needs. For example, the first upper stages that take a satellite to geostationary orbit, Centaur and the Interim Upper Stage (IUS), were designed when government contracted only for the big government and commercial satellites; McDonnell Douglas, in the late 1970s, responded by designing privately its own upper stage (the Spinning Solid Upper Stage (SSUS), also known as Payload Assist Module (PAM), which could take smaller commercial payloads into geosynchronous orbit. The company had identified a market gap and filled it with the first privately instigated and conducted space vehicle.

The Shuttle

In this new era of space transportation, the Space Shuttle is the flagship and the symbol for the U.S. space fleet. The Shuttle remains conceptually different from other spaceships in the 1980s in its reusability and versatility.

The proposed fleet of four Shuttles provides the users of space with a versatile transportation system. The Columbia, Challenger, Discovery, and Atlantis are the reusable orbiter (cargo and inhabited) portion of the Shuttle configuration. Each orbiter weighs about 75 tons and is rated for 100 missions over a decade's period. The orbiters can carry between two and seven persons easily. Its 60 foot by 15 foot cargo bay can carry 32,000 pounds to a low earth polar orbit and 65,000 to equatorial (lower latitudes) orbits.

The Shuttle has other components that augment its usefulness. The remote manipulator system (RMS) arm, constructed by the Canadian company Spar Aerospace Ltd., permits the Shuttle to lift a satellite and its upper stage booster out of the cargo bay so that it can be launched away from the Shuttle. The arm also permits the Shuttle great versatility as a repair vehicle and a transport van. Similarly, the manned maneuvering unit (MMU) permits astronauts to propel themselves in space, outside of the Shuttle, transforming the old space walk into today's space work.

Since the late 1970s, because of innovations, low earth orbit satellites have generally been designed for retrieval and repair. The Landsat D (remote sensing) and the Solar Maximum Satellite (science) both have malfunctioned in space. Solar Maximum was dramatically retrieved and repaired by the Shuttle in April 1984. Replacing the satellite would have cost upwards of $235 million; the repair mission was accomplished for $50 million. Thus, this $185 million is the first dividend of the new philosophy for low earth orbit satellites.

In addition to repair, retrieval opens up an entire new usage. The Space Telescope, which will be orbited by the Shuttle in 1987, will have a 15-year life because the Shuttle will refurbish it in space every 2.5 years and return it to earth for fuller repairs every 5 years. This procedure may also be applied to the industrial development of near-earth space. Several U.S. companies and at least one foreign competitor have begun to design retrievable space structures.

The Shuttle's versatility and its size make it the only Western spacecraft that can launch the heavy satellites, space stations, and other space objects that require human attention. The Shuttle has a present monopoly on those services, but it faces heavy competition at home and abroad for other segments of the launch market.

Expendable Launch Vehicles

The three major U.S. ELVs are the Titan, the Atlas-Centaur, and the Delta. Before the Shuttle, the Atlas-Centaur launched the largest U.S. satellites into orbit. Table 4-1 depicts the payload capacities of the other expendable launch vehicles.

The Shuttle was planned to phase out these expendable rockets. The Shuttle program, however, was delayed three or four years and did not launch until 1981. In the interim, NASA decided not to discontinue the expendable option. The decision by NASA in late 1982 to phase out expendables after all caught many experts by surprise. General James Abrahamson, then NASA Associate Administrator for Space Flight, decided that the refitting of a satellite for launch on an expendable was more expensive and time-consuming than waiting for the next available Shuttle flight.

Abrahamson's announcement generated a fury of activity in the aerospace community. The expendables had established a proven two-decade success rate of well over 90 percent. The manufacturers of these rockets—General Dynamics, Martin Marietta, and McDonnel Douglas—had some vested interest in keeping the assembly line and the launch orders coming in.

Table 4-1. Vehicles and Payloads, 1983.

Nation	Vehicle	Manufacturers	Weight to Orbit (lb)	Weight to Escape (lb) Velocity
		Initial Staging		
United States	Shuttle	Rockwell, others	65,000 (200 miles)	
	Atlas G/Centaur	GD/Convair	5,200 (geosynchronous orbit)	3,500
	Titan	Martin Marietta (UTC, Aerojet)	8,000 to 33,000 (100 miles)	2,650
	Delta 3914/3924	McDonnell Douglas	2,000 to 2,800 (synchronous transfer orbit)	1,390–2,000
	Scout	Vought	400	75
Europe	Ariane	Arianespace, Aerospatiale, ERNO	3,858 (transfer orbit)	1,750
Japan	N-2	Mitsubishi Heavy Industries, Nissan Motors	800 (geosynchronous)	770
	H-1A	Nissan Motors	1,120 (geosynchronous)	
India	SLV-3		80	

Upperstaging—United States

Stage	Fuel	Company	Pounds to Geosynchronous Orbit or Escape from Earch Orbit	
IUS	Solid	Boeing	(With Shuttle)	3,300 (planetary) 11,000 (escape)
PAM[a]	Solid	McDonnell Douglas	(With Shuttle or Delta)	1,630 (Delta class) 1,845 (Delta II class) 2,500 (Atlas Centaur class)
Centaur	Liquid	General Dynamics– Convair	(With Titan) (With Atlas)	13,000 3,500

a. PAMs (Payload Assist Modules) are available in three types to serve as an upper stage for the Delta ELV (expendable launch vehicle) or the Shuttle.

Source: Based on table in *Aviation Week and Space Technology*, March 14, 1983, pp. 144–45.

Several small aerospace firms, sometimes bolstered by venture capital, attempted to negotiate for the right to market the rockets to satellite customers. Initially, Space Services, Inc., of America (SSIA) negotiated with General Dynamics for the rights to market the Atlas-Centaur. Transpace Carriers negotiated for the rights to the Delta from McDonnell Douglas; the head of this Transpace effort had directed the NASA Delta Program in the 1970s. Meanwhile, the Space Transportation Company sought to obtain the Titan contract from Martin Marietta.

The bargaining intensified throughout 1983. SSIA pulled out of the negotiations when it became apparent that General Dynamics would develop its own subsidiary (Commspace) to market the craft. It was later reported that General Dynamics would operate the Atlas through its space programs department. SSIA then pursued the Delta option, competing with Transpace Carriers in its negotiations with McDonnell Douglas. Space Transportation Company sold both its name and its limited rights to negotiate with Martin Marietta to the Federal Express Company. Fedex Space Trans then negotiated unsuccessfully until its time limit had expired. Fedex Space Trans, however, continues to exist and has petitioned the Federal Communications Commission to enter the electronic mail/data delivery service by satellite.

In May 1983 President Reagan announced the executive decision to sell NASA's spare rocket parts and to lease the launch pads to private enterprise.[1] NASA, at the end of 1983, announced the rules for licensing and leasing space objects. And the president authorized the Department of Transportation to license private launching companies. Congress had been thought to favor placing this authority in the Department of Commerce, so a political battle may develop on this decision.

Privately Developed Launchers

In response to this billion dollar market in space transportation, more than a dozen private efforts were begun by the early 1980s. As depicted in Table 4-2, these private endeavors have centered around three basic areas—the expendables, the astronaut-rated rocket or minishuttle, and the Shuttle itself.

Table 4-2. Private Space Vehicles.

Company	Purpose	State
Starstruck (formerly Arctechnologies)	Expendables	California
Commercial Cargo Spaceline	Shuttle payloads	Washington, D.C.
Astro Tech International (formerly Cyprus Corporation)	Orbit transfer (Delta-based) fifth orbiter bid	Maryland
Earth Space Transportation System	Integrated system of Shuttle and other craft	New York
Fedex Space Transportation	Titan bid (cancelled); undecided	Tennessee
Orbital Sciences Corporation	Orbital transfer stage	Illinois
Pacific American Launch Services	Expendables	California
Phoenix Engineering	Expendables, astronaut-rated	California
Space Projects Corporation	Bid for fifth Shuttle orbiter	New Jersey
Satellite Propulsion	Expendables (cancelled)	California
Space Services, Inc., of America	Expendables—Conestoga, Delta Bid	Texas
Space Transport	Expendables	California
Stiennon Partners	Expendables, Polar	California
Transpace Carriers	Delta Bid	Washington, D.C.
Transpace	Mini-shuttle, astronaut-rated	Washington, D.C.
Truax Engineering	Expendables, astronaut-rated	California

Note: "Bid" refers to an effort by a company to obtain the contract/license from NASA to market the ELV or Shuttle commercially.

Two of the oldest efforts to develop private launch vehicles were Robert Truax's Project Private Enterprise (Truax Engineering) and Len Cormier's Transpace, both projects in the planning since the late 1960s. Truax has been developing a "Volksrocket" to carry a human being on its suborbital flight, and someone has put up $100,000 to be the first astronaut.

Transpace began in the late 1960s. Engineer Len Cormier designed a "Space Van" as a minishuttle that would be launched at 40,000 feet from the back of a Boeing 747. The minishuttle could land on any 10,000 foot long runway and would have engines for a powered descent. In 1980 Transpace filed with the Securities Exchange Commission for a capital stock offering.

Also among the pioneers of private space launchers is Gary Hudson, whose GCH, has been in existence for about a decade. He was involved in the unsuccessful testing of SSIA's first rocket, the Percheron. His Pacific American Launch Services is seeking $100 million to develop an operational liquid fuel rocket. The rocket will be reusable and may be refueled in orbit.

Each company hopes to play on a design improvement on existing space vehicles or to target a specific segment of the market in order to develop a viable company. Stiennon Partners, for instance, hopes to develop a two-stage rocket that can put a 500-1,000 pound payload into a low polar orbit. With a target date of 1986, Stiennon obviously is aiming at the remote sensing market—for contracts from the oil or ore companies (see Chapter 4).

The Starstruck Corporation has completed early testing on a water-launched vehicle. The rocket, the Dolphin, would be hauled out to sea in a transport; it would be positioned vertically with ballast and sea anchors to prepare for a relatively cheap sea launch. The first launch in 1984 was unambitious but successful. Financially, however, the company is struggling.

Space Transport has designed a new "mixed-mode" rocket engine that uses three propellants instead of two. The increased efficiency of this engine should theoretically make feasible single-stage to orbit rockets using present engineering techniques. Pacific America's Phoenix rocket already has this mixed-mode engine in its design. Space Transport is seeking the $300 million necessary to develop this rocket.[2]

The most celebrated private space launch enterprise, however, is SSIA. The Houston company received a lot of bad publicity in 1981 when its GCH liquid-fueled Percheron rocket exploded on its Matagorda Island launch pad.

After this failure, SSIA contracted with Space Vectors, Inc. to launch the rockets using solid fuel engines. SSIA also hired Eagle Engineering, also of Houston to handle design and engineering; the company hired former Mercury astronaut Donald "Deke" Slayton as president.

By late 1982 SSIA had raised more than $6 million from almost sixty investors. The company acquired a Minuteman Motor from NASA for $365,000. The motor was used in the Conestoga I, which was launched on September 1, 1982. The Conestoga reached an apogee of 195 miles and plunged into the Gulf of Mexico 260 miles away—a total success.

On the strength of this mission, SSIA began to raise more money. The company announced the construction of a new launch site on Cat Island, Mississippi. This site will allow a southern launch into polar orbit; the expended stages will drop harmlessly into the Gulf and the Pacific.

SSIA became involved in a remote sensing consortium—Space America (see Chapter 6). It has bid for the Delta expendable, but most of its interests have been focused on developing a cheap launch capacity with quick turnaround time tailored to smaller businesses.[3]

The Private Shuttle Option

In addition to Cormier's minishuttle, several offers have been made to operate the U.S. Space Shuttle as a private enterprise. William Good, an airline pilot and management innovator, founded Earth Space Transportation Systems in the 1970s to explore concepts of a Federal Express-like idea linking space to air, sea, and land. (This concept was a half decade before Fedex Space Trans seemingly adopted it.) In retrospect, Good's project, which was envisioned before the Shuttle had even been tested, had no chance to obtain the venture capital necessary for the project. In the late 1970s, Boeing also studied the possibilities of taking over commercial operations of the Shuttle.

A few years later, the idea resurfaced in altered form in the Space Transportation Company (STC) then headed by investor William Sword and economist Klaus Heiss. In 1982 STC offered to buy the fifth orbiter and lease it back to NASA, if NASA would permit STC to market the Shuttle fleet to the private users.

STC offered NASA $1 billion for the orbiter. Although it eventually decided not to follow through, Prudential Insurance Company was rumored to be seeking to purchase 40 percent of STC's stock. Later, STC began to seek the rights to market the Titan expendable. But in 1983 Federal Express bought the rights to negotiate with Martin Marietta for the marketing rights to the Titan and it also bought the rights to the name. STC became Space Projects Corporation and later Space Enterprises Corporation. Moreover, Klaus Heiss left the company in a managerial dispute.[4]

In late 1983 a new company Commercial Cargo (C^2) Spacelines, developed a new tactic to privatize the Shuttle. A pricing quirk makes it cheaper by one-third to purchase an entire shuttle flight than to buy the cargo area for smaller payloads. C^2 Spacelines plans to purchase entire payloads and then make the profit that NASA may have made. NASA might approve this arrangement because the cargo still goes by the Shuttle rather than by the foreign competition. C^2 Spacelines can arrange advantageous insurance and other technical and legal arrangements to lure more business to the Shuttle than NASA could and thereby free NASA from the mechanical task. In fact, the plan is actually an incremental version of the old STC plan, which would have marketed the entire shuttle schedule privately.[5]

Although NASA and the Reagan administration are very tempted by aspects of the privatization of the Shuttle, several factors have dictated caution. NASA has long had an internal dispute as to whether it should be a research and development agency or become an operational organization. If NASA must operate the Shuttle, the organization must become more business-oriented and less research-oriented. Yet, as space becomes more of an arena of international competition, research and development must be continued for the United States to keep its competitive advantage. This conflict in NASA between research and operations has never been resolved.

The U.S. government also has been a notoriously bad entrepreneur. For instance, NASA is not permitted to build a fifth orbiter because the Reagan administration has decided there is competition for the market! The government has been a poor business decision-maker because other factors enter in decisions. Government policy is based on partisan and interest group politics instead of on business or technological grounds.

Nonetheless, McKinsey and Company, management consultants, in the late 1970s, and the National Association of Public Administration, in the early 1980s, held that NASA should retain the Shuttle at least until it had become a proven operation. The Shuttle had too many national security and other interests to be risked in a private operation. It would seem also that a fleet of only four would be too big a risk—losing one shuttle in an accident would be a loss that a private enterprise simply could not absorb.[6]

U.S. Upper Stages

A similar proliferation of orbital upper stages seems to be in the offing. Until the 1980s, upper stages were government designed and contracted. These boosters were designed for large payloads, but they were victims of a series of delays by politics and engineering. The two competitors for NASA/Air Force contracts were Boeing's Interim Upper Stage (IUS) and the Centaur for both the Shuttle and the ELVs.

Both upper stages, however, were expensive and more powerful than needed by many of the new commercial satellites. Boeing and General Dynamics, who makes the Centaur, became involved in a major lobbying battle over which stage would be funded at what rate. Politics delayed both crafts. In light of this situation, McDonnell Douglas developed its own Spinning Solid Upper Stage (SSUS). These payload assist models (PAMs) were produced in three types that could be used to launch smaller payloads from a Thor-Delta or the Shuttle. The PAMs were successful in their first eighteen missions until the publicized double failure on the Shuttle in early 1984.

In the early 1980s, several graduate students earned their Harvard MBAs with a plan to develop a private orbital upper stage. All began well. These graduates organized a new company—Orbital Systems Corporation, later Orbital Science Corporation (OSC)—and worked with Martin Marietta Corporation on initial plans. In 1983 NASA discontinued its own upper stage design and began to negotiate with OSC for their Transfer Orbit Stage (TOS).

This TOS would have less technical electronics, but its liquid fuel rockets would permit it to put 6,800 pounds in geosynchronous, compared with 5,000 for the IUS and 5,200 for Ariane 4. Moreover, TOS (with the shuttle) would cost $81.6 million compared to IUSS's $125 million (with the shuttle) and Ariane 4's $75.4 million.

But OSC, for all its detailed academic planning, found itself in a changed and threatening situation by late 1983. The earlier market studies had relied on the new direct broadcast satellite industry; this prognostication had been based on large, high-powered satellites, but the technology is now permitting low-powered smaller direct broadcast satellites that can be launched from the PAM-D. The newer model satellites also will be equipped with their own liquid-fueled propulsion systems. OSC was recently bypassed as Hughes Aircraft purchased its Intelsat VI solid motors directly from United Technologies. Moreover, the European large payloads are likely to be launched on the Ariane, and the U.S. interplanetary program, which could use TOS, remains in disarray.

Although OSC may yet be successful, it shows the rigors of a small entrepreneur in a new industry. Market trends in a stable industry are difficult to forecast, but where the technology is just developing and the market and customers are just aborning, market forecasting here is the equivalent of drawing for an inside straight. And when it is a small, speculative company, the deal is for all the chips.

Later at the end of 1983, Astro Tech International began negotiations to acquire parts of the Delta Upper Stage. This upper stage could place payloads weighing 5,000 to 20,000 pounds into geosynchronous orbit.[7] The market and competitors are nascent and in flux.

FUTURE VEHICLES AND SPACE INFRASTRUCTURE

In the next century, the space around the earth will be crowded with familiar satellites and shuttles as well as several new crafts and structures. Heavy lift vehicles (HLVs) will be able to carry cargos larger than the Shuttle's into orbit. These cargoes will include huge satellites and even the components for large space structures and space stations.

Some of these space bases will remain in low earth orbit. Others will be ferried to geosynchronous orbits or even beyond to the L-points. (The L, or Lagrangian points are the positions in the earth/moon/sun system where the gravitation effects of the three bodies are balanced and anything (including a space base) put there stays there.) These structures in orbit may be large communications satellites, solar-powered satellites, space stations or colonies in space.

Each of the objects will require a host of services in order to operate smoothly. Private industy will probably be able to identify the goods or services that will be needed. The resourcefulness of earth-based and now space-based private enterprise suggests no less. Yet, the experience in this first generation of space entrepreneurship suggests some caution. Jumping into a market before it has solidified is the speculators' urge. Although some of these new companies will be successful and one may be the stock market's IBM of the 1990s, most new space ventures are destined to noble failure.

FOREIGN COMPETITION

Competition, from Europe and potentially from Japan and the Soviet Union, suggests that the United States must carefully guard this space market, one of the last areas in which the United States holds a commanding lead in technology and expertise. The United States must deal with this on several levels—marketing, politics, and engineering.

The United States must decide how to coordinate its public and the private space efforts. The two may conflict; for instance, if the United States pursues a pricing policy for the Shuttle to compete with the Ariane, the U.S. private market for expendable rockets may also suffer. If the United States engaged in price cutting, Ariane, with European national backing, will be able to withstand the pressure. The U.S. private companies may not.

Since the United States seems committed to a four-Shuttle fleet, a low-priced launch should generate new business that would probably go to the European launchers. Although Ariane has only two launching pads, Arianespace could increase production of its expendable rockets and build new sites much cheaper than the U.S. Shuttle fleet and capacity can be expanded. NASA and the government, however, can subsidize the U.S. private launchers by several financing techniques.

The United States is permitting these companies to lease launch facilities and to buy spare parts. The tax laws are already written to encourage high-tech endeavors. Insurance to protect both the company and the nation (under treaty law, the United States is responsible for any damage caused by its nationals in space) has become more difficult to obtain because of recent failures in the communications satellite industries. At present, NASA does provide insurance for its payload, but this insurance does not extend to the new area of private launching. It would not be hard to extend these provisions to launchers using NASA facilities, but many of the new companies plan to launch from other places, some from beyond U.S. territory (although the United States is still bound by treaty to regulate its nationals anywhere). (See Appendix B for the international convention on space liability.)

Several bills have been offered in Congress to remedy the dual problem of U.S. treaty obligation and protection of the nascent industry. The bills have generally called for either the Federal Aviation Administration (FAA) or the Department of Commerce to license and otherwise regulate the private launchers. Such a bill would regularize the industry; insurance provisions could be included. Establishing one lead agency, moreover, would remedy a problem already faced by SSIA in its efforts to launch the Conestoga.

The company needed clearance from almost a dozen federal agencies to launch the Conestoga. It had to receive clearance from NASA, the FCC, the FAA, the North American Air Defense Command (NORAD), the State Department, the Defense Department, and even the Alcohol, Tobacco, and Firearms Administration (because a rocket is considered an explosive device). The State Department required the company to get an export license even though the rocket would land in the Gulf of Mexico. The FCC granted the company a radio frequency and an experimental license, although this procedure would not be appropriate once such a company became operational. But the bottom line is that any one of these agencies could have prevented the launch, or two agencies could have introduced contradictory requirements.[8]

Congress, therefore, will have to designate one lead agency for space launchers in the near future. The FAA seems on the surface to be the best organization because it already regulates the air lanes that spacecraft must transverse; the FAA also has experience with safety regulations.

If space enterprises were limited to launch vehicles, the FAA should be the lead agency. The lead agency, however, will also need to regulate the payloads of these vessels, especially as space manufacturing becomes more and more common. The Commerce Department or a new Department of Space is the most appropriate solution to this problem. (See Chapter 9, on space policy.) In another view, the establishment of a clear regulatory scheme for launching enterprises marks a continuing maturation of space commerce, similar in importance to the role of the FCC in regulating the burgeoning communications satellite industry.

In November 1983 President Reagan signed an executive order authorizing the Department of Transportation to license and otherwise regulate private launch vehicles. The Office of Commercial Space Transportation in the Department of Transportation came into existence in 1984. Influential members of Congress, however, continue to favor the Department of Commerce as the better seat for a coordination of space regulatory activity. (See Appendix H for the text of the bill.) A "Space Commerce Act" would provide the needed supervision for the private space launchers, but it does not deal adequately with the broader questions of space commerce. The House conducted hearings on HR 3942, a bill that would reinforce the president's grant of authority to the new office.

NOTES TO CHAPTER 4

1. *NASA Office of Space Flight*, 1983; *Space Calendar*, May 16-23, 1983, May 23-29, 1983; *Austin American Statesman*, May 18, 1983; *Commercial Space Report*, December 1983.
2. *Commercial Space Report*, November 1983; *Space Business News*, August 15, 1983; *Space Calendar* and *Commercial Space Report* provide an ongoing account of these new space businesses.
3. *Science News*, September 18, 1982; *Time*, September 20, 1982; *Austin American Statesman*, January 17, 1982, September 12, 1982, October 12, 1982; *Aviation Week and Space Technology*, September 13, 1982, June 14, 1982, September 6, 1982; *Houston Post*, September 21, 1982.
4. *Satellite Communication*, August 1982, January 1983; *Aviation Week and Space Technology*, July 19, 1982, August 9, 1982, September 13, 1982, November 29, 1982; *Space Calendar*, July 11-17, 1983.
5. *Commercial Space Report*, December 1983.
6. McKinsey and Company, Inc., *Organizing for Effective Operation of the Space Transportation System* (NASA: Lyndon Baines Johnson Space Center, May 1980); National Academy of Public Administration, *Encouraging Business Ventures in Space Technologies*, prepared for NASA, 1983.
7. *Commercial Space Report*, November 1983, December 1983; Orbital Systems Corporation, *Executive Summary*, February 1, 1983.

8. James R. Myers, "Emerging Government Regulation of American Space Entrepreneurs," in *L-5 Society National Conference: Doing Business in Outer Space*, Houston, Texas, April 1983.

5 TELECOMMUNICATIONS

In the late 1950s, rockets were conceived as the conveyors of warheads and warriors. But very quickly, engineers and scientists found other uses for the rocket. Communications and remote sensing (surveillance) became the first space applications.[1]

Although they seem different, these two applications employ much similar hardware and technology, have similar histories and raise many of the same questions of policy and law. Both systems require ground stations to emit and receive signals from the satellites; the data in the signals are digitally coded to be stored on computer. But the communications industry has matured faster than remote sensing, becoming by the 1980s a multibillion dollar endeavor. Remote sensing has yet to find its mature market. Nonetheless, both technologies are the focus of international law disputes over seemingly incompatible doctrines of national sovereignty and of free access to information across borders.[2]

From the beginning, both applications technologies were employed by the civilian and the military space authorities. Communications and remote sensing could serve both masters. The military headquarters, with instantaneous, reliable worldwide communications, can coordinate and direct activities in the field, effecting what is called a "force multiplication" of available troops. Similarly, better weather forecasting as well as surveillance of enemy troops can at least prevent nasty surprises on the battlefield. The coordination of these technologies facilitates the military's C3I—communications, command, control, and intelligence—and plays a major role in the modern balance of power between the United States and the Soviet Union.

COMMUNICATIONS HISTORY AND HARDWARE

As early as 1958, the U.S. Score Satellite, launched by the military, demonstrated the potential of space-based communications by broadcasting a taped Christmas message from President Eisenhower to the world. The NASA balloon, Echo I, in August 1960, succeeded in relaying radio signals across intercontinental distances to large receivers. America's technical prowess in communications satellites advanced rapidly.

Soon after the success of Echo I, American Telephone and Telegraph (AT&T) inquired whether NASA would consider providing launch services at cost, so that AT&T could investigate the commercial potential of space-based communications. The Eisenhower administration approved the concept in the hopes of maintaining or acquiring U.S. leadership in space and space applications.

AT&T then proceeded with its low-altitude Telstar I satellite. This spacecraft was the first active satellite with a transmitter. In 1962 Telstar transmitted the first live transatlantic television. Telstar I failed in 1963 because of radiation damage, but Telstar II proved further the potential of space communications as a commercial market.

These first satellites were placed into low earth orbits. TV or other reception, therefore, could only occur when the satellite was overhead or in the line of sight. One shotgun approach to continuous communications would be to stagger a series of low earth orbital satellites so that when one had gone beyond the range of communications, a second satellite in the relay could continue the broadcasts. Still, gaps in reception could come at inopportune times, such as during U.S. broadcasts of European news or sporting events.

Hughes Aircraft Company began about the same time to develop its own *Syncom* satellite series. Syncom I, launched in February 1963, reached geosynchronous orbit but failed to operate. Syncom II, in July 1963, and Syncom III, in August 1964, proved the concept of communications from geosynchronous orbit.[3]

With the development of larger rockets and more powerful satellites, the modern solution for instant and inclusive communications came into being—the geostationary satellite. These satellites, positioned 22,300 miles above the equator, match speed with the earth so that they are always located over the same spot on the equator.

Because there is a limit to the geostationary points (360° of the circle), these points or slots are at a premium. These slots are viewed, therefore, as a limited, precious resource; the International Telecommunications Union (ITU) decided in a series of meetings in the late 1970s to the mid-1980s to allocate these slots among the nations of the world.

The United States and other space powers have pursued a second strategy for using these orbital slots and radio spectrum. They have advocated an open evolutionary access to these slots and improved technology to use the resource better. One strategy has been to increase the frequencies or bandwidths used for broadcasts. All early satellites broadcasted in the 6/4 gigahertz (GHz) or C-band. The C-band is the wavelength of microwave so that satellites do not need equipment to convert the wavelengths for transmission. Six refers to the uplink; that is, the radio signal is beamed to the satellite at 6 GHz and returned to earth at 4 GHz.

By the 1970s, it was evident that the 6/4 range was filling up and that new bands had to be opened up. 14/12 GHz, the Ku-band, was developed. This frequency does require more complex equipment for transmission and reception and is more attenuated in bad weather; nonetheless, this frequency is also filling up quickly. One of the major policy questions of the late 1980s will be research and development (R&D) funding for even higher frequencies such as 30/20 or 50/40 GHz already included in U.S., Japanese, and European plans. These higher frequencies, however, require technical advances so that they can penetrate the atmosphere.

Each satellite has a number of receivers, called *transponders*, which receive the signal from the ground and amplify and return it to another earth station. Transponders have traditionally been tuned to one frequency bandwidth and also traditionally one bandwidth per satellite. (The trend has been continually to more transponders, often with several bandwidths and with increasing ability to switch frequencies on the space satellites.) Moreover, spot beams that can broadcast to specific areas and higher powered satellites that can broadcast to small earth receivers have further increased the versatility and the market for these satellites.

THE INTERNATIONAL SPACE BUSINESS

Communications by satellites have become the first truly worldwide major space business. The availability of satellites to fill communications needs not fulfillable with ground-based alternatives, at costs that allow profit, has engendered a multitude of domestic and international organizations in space-based communications. These organizations include private domestic corporations, national consortia, multinational corporations and international governmental organizations. By 1983, 6.7 percent of the communications assets worldwide were space-based, that is, $10 billion of the $150 billion market. By the year 2000, the space-based segments are expected to rise to 9 percent—77 billion of the $850 billion investment.[4]

Comsat

At the same time of AT&T's and Hughes Aircraft's pioneering efforts, the Congress undertook other steps to assure that both the United States and private enterprise would be represented in space. The Communications Satellite Act of 1962 was partially in response to the international situation. The Soviet Union was attempting in its political propaganda to equate private enterprise and piracy in an effort to ban private activity in space by the emerging international law of space. The Communications Satellite Act, creating *Comsat* as a public/private company, presented the world with a fait accompli—private enterprise in space.

NASA supported Comsat with both R&D and launch services. According to the Comsat Act, Comsat would be managed by a fifteen-member board of directors. As a public/private operation, twelve directors were chosen by the shareholders and the other three were appointed by the U.S. president and confirmed by the Senate.[5]

Comsat first issued stock to the public in 1964. In April of 1965 it paid NASA to launch its first commercial satellite, Early Bird, into synchronous orbit.

Two years later Comsat showed its first quarterly net operating profit and in 1970, it issued its first quarterly dividend. By 1980 Comsat had 80,000 shareholders holding over $40 million worth of stock. Comsat also holds several subsidiaries and is involved in varied joint ventures. While Comsat General Corporation operates the domestic system and operates R&D satellites, Comsat General Telesystems manufactures electronics and other components, and Environmental Research and Technology builds the equipment to monitor the environment. Comsat's Satellite Television Corporation was created to proceed with direct broadcasting technology.

Comsat has also been involved in several joint endeavors. Along with IBM and Aetna Life Insurance Company, Comsat is a shareholder in Satellite Business Systems (SBS)—a company established to fulfill the communications needs of major corporations. SBS has become successful whetting Comsat's interest in other space ventures. In 1983 Comsat joined with a German company (MBB) and a Swedish investment group (Stenbeck) to create Sparx Company to commercialize remote sensing with the German Shuttle Pallet Satellite—SPAS (see Chapter 6).[6] Comsat's reorganizations, however, have refocused the company's ventures.

Intelsat

Comsat became the U.S. representative and first manager of the International Telecommunications Satellite Consortium (*Intelsat*) founded in 1964. Since 1980 Intelsat has managed its own programs, although Comsat is often contracted to operate its new programs. Intelsat now has more than 100 member nations and owns and operates a series of satellites that control most communications between nations; a few nations even use Intelsat for their own domestic systems. The ground stations, however, are owned by the individual nations or by private corporations.

Intelsat represented several U.S. policy victories. The organization permits nations to invest in, share profits, and vote (on some matters) according to its input into the consortium. Moreover, private corporations can represent a nation within Intelsat. In addition to the U.S. representative, Comsat, private companies also represent the communications interests of Portugal, Italy, and several other nations.

Structurally Intelsat is composed of four administrative bodies.

1. The Assembly of Parties meets on a one-nation, one-vote basis on issues concerning sovereignty.
2. The Meeting of Signatories includes the representatives from the national signatory. A nation's signatory may be a government agency or, as in the case of Comsat, a private corporation.
3. The Board of Governors handles managerial decisions and decides those issues based on the weighted vote of the members. Members include the largest shareholders, such as the United States and the United Kingdom and regional blocs of pooled shares. The Board presently has an African, a Caribbean, and a Nordic nloc.
4. The Director General operates a Washington-based Executive Organization and is responsible to the Board of Governors.

Intelsat was so successful politically and financially that maritime nations initiated the International Maritime Satellite Organization (*Inmarsat*). By 1983 over three dozen nations had joined, including the United States and the Soviet Union as its two largest shareholders. This system will link thousands of ships to the shore and to each other with relatively little atmospheric interference. Tables 5-1 and 5-2 show the distribution of shares in Intelsat and Inmarsat, respectively.

The decision by the Soviet Union to join Inmarsat represents a break with its political past. The Soviet Union had refused to join Intelsat because of the view that voting based on shares rather than on statehood violated national equality and sovereignty. The Soviet bloc presently operates Intersputnik as a rival (one-nation, one-vote) organization to Intelsat. Whether this shift means that eventually Intersputnik and Intelsat will merge cannot be determined at this time.

Other Regional and National Organizations

Arabsat is representative of several new regional telecommunications organizations. Twenty Arab countries and the Palestinian Liberation Organization (PLO) comprise its membership; it became operational in 1984. The European Telecommunications Satellite Organization, *Eutelsat* also became operational in 1984. Twenty nations (mostly in Europe) own the company, which uses the ECS satellite launched by

Table 5-1. Intelsat—National Shares.

Country	Signatory	Investment Share (%)
United States	Comsat	22.6
United Kingdom	Post Office	12.9
France	Government	6.5
Federal Republic of Germany	Federal Ministry for Post and Telecommunications	3.4
Brazil	EMBRATEL	3.2
Japan	Kokusai Denshin Denwa Company	3.1
Saudi Arabia	Government	3.0
Italy	Societa Telespazio	2.8
Australia	Overseas Telecommunications Commission	2.7
Canada	Teleglobe Canada	2.5
Nigeria	Nigerian External Telecommunications Limited	2.1

Source: Data from Comsat, Seventeenth Annual Report to the President and the Congress as of September 30, 1980.

Arianespace. *Nordsat* serves the Scandinavian region; *Palapa*, owned by Indonesia, serves much of Southeast Asia. Subsaharan Africa and South America have also begun discussions to establish regional satellite systems.[7]

In addition, many countries have their own satellites. The French and Germans cooperated on the Symphonie satellite in the 1970s; and the TDF-1 and the OTV satellite in the 1980s. Arianespace will launch Italy's *Italsat* in the 1980s.[8] Japan relied on the Telecommunications Satellite Corporation of Japan, Telesat-Japan (co-owned by Kokusai Denshin Denwa) to conduct Japan's Intelsat and Inmarsat business. Japan has taken the world lead in direct broadcasting at higher frequency with experimental craft. In 1985 Japan will launch its YURI satellites to provide such a national communications network.[9]

Since the middle 1970s India has been involved in building or buying communications satellites. India's new satellite, *Insat*, which can

Table 5-2. Inmarsat Member Nations and Their Percentage Investment Shares.

Nation	Share (%)
United States	23.4
Union of Soviet Socialist Republics	14.1
United Kingdom	9.9
Norway	7.9
Japan	7.0
Italy	3.4
France	2.9
Federal Republic of Germany	2.9
Greece	2.9
Netherlands	2.9
Canada	2.6
Kuwait	2.0
Spain	2.0
Sweden	1.9
Denmark	1.7
Australia	1.7
India	1.7
Brazil	1.7
Poland	1.7
Singapore	1.7
People's Republic of China	1.2
Belgium	0.6
Finland	0.6
Argentina	0.6
New Zealand	0.4
Bulgaria	0.3
Portugal	0.2
Algeria	0.1
Egypt	0.1
Iraq	0.1
Liberia	0.1
Oman	0.1

Source: Data from Comsat, Seventeenth Annual Report to the President and the Congress as of September 30, 1980.

combine remote sensing and communications (including direct broadcast), was constructed by Ford Aerospace. The Germans have been helping China with its satellite programs. Brazil also has received foreign assistance in developing its program. Mexico, Colombia, and other developing nations likewise have long-term plans to build or buy satellites to provide communications. U.S. and European companies will design or build most of these early efforts.[10]

Table 5-3 depicts the depth of the international market in space communications. The table also suggests the competitiveness for both the satellite construction and the launching contracts.

The U.S. Industry

Today both the market in communications satellites and the competition for satellite contracts are worldwide. Again the United States was the first to pioneer both the uses and the marketing of satellite communications.

Between 1970 and 1972, the Federal Communications Commission authorized the first domestic communications satellites. These satellites soon became the first space industry to exist independently of the government infusion of funds. First telephone communications and then television provided the revenues for this industry.

It was the "Thriller from Manilla," boxing's heavyweight championship fight between Muhammad Ali and Joe Frazier, September 30, 1975, that proved that Time's Home Box Office (HBO) and subscription television had a future.[11] Private industry quickly began to develop new uses and new services for the communication satellites. Until the middle 1970s, satellites' bulk business was transmittance of long-distance telephone calls and special events on live television. The 1980s witnessed the mushrooming of cable television, with a majority of the subscribers linked by satallite.

A quick review of Table 5-4 reveals several important aspects of the satellite industry. The table lists almost 80 satellites in space or planned for the late 1980s. Until the early 1980s, a scarcity of satellites and satellite transponders had characterized the industry. The FCC responded by approving in one year the locations and frequencies for more than 20 new satellite systems. The motivations of the agency were partially a response to the need for more capacity and

Table 5-3. International Satellites.

Satellite	Prime Contractor	Launcher	Number	Purpose	Bands	Number of Transponders
Intelsat V	Ford	U.S.	9	International telephony, TV, teletypewriter, data, some domestic uses	C, KU	25, 6
Intelsat Va	Ford	U.S./Europe	6	International telephony, TV, teletypewriter, data, some domestic uses	C, KU	32, 6
Intelsat VI	Hughes	U.S./Europe	≤ 16	International telephony, TV, teletypewriter, data, some domestic uses	C, KU	36, 10
Marecs	British Aerospace	Europe	3	Mobile telephony, teletypewriter	L-ship C shore	1, 1
Arabsat	Aerospatiale (France)	U.S./Europe	2	Regional telephony, TV	C, S	25, 1
Palapa (Indonesia)	Hughes	U.S.	2	Regional telephony, TV	C	24
ECS (Eutelsat)	Mesh (Europe)	Europe	5	Regional telephony, TV	KU	12
Telecom (France)	Matra (France)	Europe	2	Regional telephony, TV	C, KU	6, 6
Anik C (Canada)	Hughes	U.S.	3	Telephony, telegraph, data, TV	KU	16
Anik D	Hughes	U.S.	2	Telephony, telegraph, data, TV	C	24

TELECOMMUNICATIONS 65

Aussat (Australia)	Hughes	U.S./Europe	3	Telephony, radio, data, TV	KU	15
Insat (India)	Ford	U.S.	2	Telephony, teletypewriter, data, TV, meterological	C, 5	4, 2
CS (Japan)	Mitsubishi/Ford	Japan	2	DBS, data, telephony	C, KA	2, 6
Yuri (Japan)	Toshiba/G.E.	Japan	2	TV direct broadcast (DBS)	KU	2
OTS–2	Mesh/BA/Telefunken	U.S.	1	Experimental European communications	KU	6
TDF–1	Aerospatiale, MBB	U.S.	?	DBS, FR–Luxembourg–Germany	KU	?
Mexico	Hughes	U.S.	2	TV, telephony, data	C, KU	24
Tele-X (Sweden)	Aerospatiale/Europe Satellite	Europe	?	DBS, video data transmission	KU	?
TV-Sat	?	?	?	German DBS	KU	?
Olympus	British Aerospace	?	?	ESA DBS	KU, KA	?
Unisat	United Satellite Ltd.	?	?	UK–DBS	KU	?

Source: Based on data in *Electronics*, October 6, 1982; *Aviation Week and Space Technology*, March 14, 1983; and other sources.

Table 5-4. U.S. Communications Satellites.

Satellite	Prime Contractor	Numbers Planned	Purpose	Bands	Transponders
Westar (Western Union)[a]	TRW	6	Telephony, telegraph, TV	C, Ku	4, 12
G-Star (GTE)	RCA	3	Data, DBS	Ku	16
Spacenet (SPCS)	RCA	4	Telephony, telegraph, data, TV	C, Ku	18, 6
Galaxy	Hughes	4	Telephony, data, TV	C	24
Telestar-3 (AT&T)	Hughes	3	Telephony, telegraph, TV	C	24
Fordsat	Ford	2	Telephony, telegraph, TV	C, Ku	24, 30
Satellite Business Systems	Hughes	3	Data, TV	K	10
ASC (Am Sat)	RCA	4	TV, voice, data	C, Ku	18, 6
USAT (U.S. Satellite System)	?	4	Telephony, data, TV	Ku	20
DSCS-2	TRW	15	Military	X	4
DSCS-3	G.E.	12	Military	S, SHF	6
NATO-3	Ford	5	Military, telephony, teletype	X	2
Navy Leasat	Hughes	4	Mobile military	UHF, SHF	9
Satellite Data Systems	Hughes	?	Air force	UHF	?
SatCom	RCA	4	Telephony, TV	C	24
Western Union	Hughes	3	Telephony, teletype, data, TVD	C	24
Marisat	Hughes	3	Naval communication	?	?
Comstar (Comsat)	Hughes	4	Leased to AT&T	C	24
Americom	RCA	3	Telephony, data, TV	C, Ku	24, 16

a. Western Union dropped out of this program and it is not clear whether the program will be undertaken.

Source: Based on data in *Electronics*, October 6, 1983; *Aviation Week and Space Technology*, March 14, 1983 and other sources.

partially a response to the Reagan administration's deregulation of the communications industry.

Table 5-4 also reveals that only a handful of corporations handle the prime contracts for the assemblage of satellites. These are major corporations in the related fields of electronics and aerospace. Hughes Aircraft is the largest producer of communications satellites. Other major prime contractors are Ford Aerospace, RCA, TRW, and General Electric. Increasingly, European and Japanese firms are gaining the expertise to build all or parts of satellites. Because satellites are so complex, many components are subcontracted to other companies—either independent concerns or the subsidiaries of the prime contractor. When the FCC approves a new series of satellites, the transaction will lead to millions of dollars in contracts for launch services and for ground control, including satellite commands and data transmission.

The communications full-service common carriers also include many well-known corporations—AT&T, GTE, RCA, and Western Union—as well as lesser known concerns, such as the Southern Pacific Satellite Communications Corporation (SPSC). Usually, these companies own their satellites. In addition to using their own satellites, they may lease or sell spare capacity on the transponders to other companies.

Another set of companies may act more indirectly as the middlemen of communications. Resale carriers lease either the transponder or time on the transponder and sell that time to companies for broadcasts. Such programming can either be one-time events or ongoing services.[12]

Since the early 1980s, the FCC has approved a new form of satellite property right, the *condosat*. With the growing market for satellite communications and the scarcity in transponders at the time, several satellite owners began to sell rather than to lease the transponders. The satellite builder receives a quick return on the investment; the user-broadcaster now can depreciate the transponder on income taxes as business property.

RCA, Western Union, and Hughes Aircraft began to auction transponders on the Satcom-5, Westar-V, and Galaxy. The FCC permitted the rates but rejected the use of the auction prices because the unequal prices violated the "public interest" standards of the 1934 Communications Act, under which the FCC regulates and all domestic communications operate. The companies then used the auction price average, $13 million, as the price per transponder. By 1984 the cost of a transponder had gone down to under $10 million.[13]

The big purchasers of these transponders have been the cable networks such as HBO, CNN, and MTV. All three have revolutionized American television, and all three depend heavily on satellites for reaching their publics. Throughout the 1970s, more and more homes in America acquired subscription television. In the 1980s, the Satellite Master Antenna Television (SMATV) began to emerge as a new subfield. SMATV companies can hook up an entire apartment complex with a pay television system; this technology is also used to set up small earth stations in backyards to receive satellite signals.

By 1984 this industry was approaching a crossroads. Several cable companies such as HBO and Showtime had decided to scramble their signals so these services could retain exclusivity of their product.[14]

The SMATV private cable operation and the dealers of backyard earth stations have begun to fight this threat in both the Congress and the courts. Two trade associations represent these two industries—the Society for Private and Commercial Earth Stations (SPACE) and the National Satellite Cable Association. SPACE has attempted to require the companies like HBO, who scramble their signals, to allow reasonable access charges to SMATV or backyard operators.[15]

In the courts as in Congress, these small companies want an environment that will permit them a free hand to develop. Courts in Minnesota and New Jersey have held that states may regulate their private cable operations. The New Jersey case has been brought to the FCC. SPACE, however, is arguing that the FCC should not preempt the state's right to regulate the market. The trade association does not want such a preemption because it fears it as a first step to national regulation.[16]

Direct Broadcasting Satellites. The SMATV, however, faces a combined threat and hope in the developing of a direct broadcasting satellite (DBS). The idea of DBS is almost two decades old; in its classic form, it would be a powerful satellite that could transmit a focused beam directly to small home receivers. The technology has improved so much that DBS has become possible with more traditional, less powerful satellites; this technology can manage DBS-like services to a small antenna from a standard 14/12 GHz satellite.

In 1983, Oak Satellite Corporation prepared to become the first DBS company in the United States when it leased four transponders on Canada's Anik C. satellite and transmitted signals at 12 GHz to 1-meter home antennas. The FCC received more than a dozen requests for DBS systems in 1982. The FCC approved nine of these systems—CBS, Comsat's STC, DBS Corporation, Focus Broadcasting, Graphic Scanning, RCA, USSB, VSS, and Western Union. Other later entries include Skyband, owned by publisher Rupert Murdoch (who later backed out of it) and United Satellite Communications (USCI), backed by the Prudential Insurance Company. USCI, however, was the company that inaugurated DBS with its service to central Indiana, with plans to expand throughout the country.

The National Association of Broadcasters (NAB) opposes DBS on at least three grounds:

1. DBS consumes too much of the bandwidth.
2. It has filled up the last remaining geosynchronous slots.
3. Its full application might endanger the economics of local broadcasting.

Although NAB lost in the FCC, NAB continues to lobby Congress to overrule the FCC and change the law.[17]

In response to the DBS and other high-powered satellite needs, both Hughes and Ford have designed new satellites. The new Hughes design will be able to operate 48 transponders. Ford Aerospace has designed its new *Supersat* as a modular craft capable of operating in the Ku, C, and L bands as well as with direct broadcast and spot beams. In the C-band range, the Supersat could maintain 34 channels. Ford plans to use this satellite initially in its own domestic communications subsidiary, Ford Aerospace Satellite Services Corporation.[18]

DBS and SMATV enter a crowded communications market already occupied by satellite cable networks and several other offerings on the horizon, such as low-power television (LPTV). The DBS and SMATV services can complement each other but could devastate each other if, for instance, DBS or cable refuses to provide reasonable access to its programming. SMATV has threatened then to begin antitrust proceedings.[19] The economic viability of all these communications options suggests that a period of experimentation will end with a shakeout, in which many companies and some services unable to compete will drop out of the running. Nonetheless, the vitality of the market likewise suggests a bright economic future for the services in the communications/entertainment field.

The Full Service Satellite for Business Communications. Satellites have greatly improved the productivity of businesses. Teleconferencing has permitted companies with offices on both coasts to conduct business directly without the expense in time and money for travel. Even though NASA had been praising satellite teleconferencing for decades, only by 1980 had the concept reached the million dollar revenue threshold. By 1985, however, the industry expects revenues of $380 million.[20]

Many major hotels, especially with convention centers, provide teleconferencing as a service for their customers. For example, Public Service Satellite Consortium coordinated over 20 teleconferences between August 1983 and March 1984. Activities as diverse as hospitals and commodity markets were using teleconferencing and other satellite services to conduct their affairs more efficiently.[21]

The hook-up of computer and satellite technologies has already begun another great change in the way we do business. The distinction between voice and data communications has been obliterated by the digital transmission of information. In addition to TV and telephone, telex or electronic print are creating several new markets. Many industries are moving away from normal phone lines to use satellites for all types of data transmissions. Satellite Business Systems was the first firm to tap into this business. It is a consortium of IBM, Comsat, and Aetna Insurance companies. But other companies have moved quickly to prevent a monopoly in the field. One reason that AT&T did not fight the divestiture of its local phone companies was its desire to rid itself of those regulated and less profitable operations. AT&T hopes to enter these unregulated services, including electronic mail and other endeavors. Similarly, MCI, GTE (Spacenet Satellites), and other long-distance services are catering to the business community.[22]

A final indication of how extensive space communications has become is the phenomena of the Teleports. The Teleports will be located outside of the cities to avoid radio interference. These sites will also be shielded to protect them further from interference. The first Teleport, located on Staten Island, will have seventeen earth stations that will permit access to all U.S. domestic satellites as well as the international. Cables of optical fibers will connect the Teleport to switching systems and then to businesses in the cities. The cables will be able to carry all services, including voice, video, data, and teleconferencing.

The New York City Teleport will be a $350 million investment including an office park and communications center. The project is funded as a joint venture of Western Union, Merrill Lynch, and the Port Authority of New York and New Jersey. Similar teleports are planned for other cities in the United States.[23]

THE FUTURE OF SATELLITE COMMUNICATIONS

The satellite communications industry has been marked by technological and market surprises. Many new miraculous services will be introduced through the rest of the decade and beyond.

New services will be provided by satellites in the mobile radio/telephone industry. This service will be especially appreciated in the rural areas. Returning to the urban scene, video teleconferencing may one day become three dimensional—that is, holographic. High-powered DBS may make the Dick Tracy watch TV/radio a reality in only a few years. Already, a new company, Geostar, headed by Gerard K. O'Neill of the Space Studies Institute and Princeton University, has filed an application with the FCC to permit a positioning/location satellite service. This service would permit a person with a pocket transmitter to beam his or her location to a satellite. Ships, aircraft, or hikers lost in difficult circumstances can be tracked and rescued.

The new high-powered satellites, and the following generation of antenna farms in space, open up new opportunities for business, society, and politics. On the beneficial side, the space age cottage industry—a person working at home tied by computer and satellite to a central office—is already a reality. The phone-in-polls of television also presage the first real opportunity for worldwide democracy through electronic voting. Yet the reverse is also a possibility with this technology. It is 1984: electronic surveillance of citizens by satellite is also a possibility.

Satellite communications is a great opportunity and a great challenge to American life. As the Congress reconsiders the Communications Act of 1934, by which the FCC regulates the industry, the lesser and the greater implications of this new technology must not be ignored. Deregulation to permit private development is important. International legal questions (see Chapter 9) are also important, but the ultimate impact of these technologies on the entire of the world's society must also be considered in U.S. space policy.

NOTES TO CHAPTER 5

1. *The National Aeronautics and Space Act of 1958.* P.L. 85-568, 85th Congress, H. R. 12575, July 29, 1958.
2. See generally, James Martin, *Communications—Satellite Systems* (Englewood Cliffs, N.J.: Prentice Hall, Inc., 1978); Delbert D. Smith, *Space Stations International Law and Policy* (Boulder, CO.: Westview Press, 1979).
3. *Satellite Communications*, October 1982.
4. "Space Station Programs: Final Study Report," Lockheed, April 22, 1983.
5. *The Communications Satellite Act of 1962*, P. L. 87-624, 87th Congress, H. R. 110-40, August 31, 1962.
6. 17th Annual Report to the President and Congress, Comsat.
7. "Multilateral Intergovernmental Cooperation in Space Activities," 2nd U.N. Conference on the Exploration and Peaceful Uses of Outer Space, January 30, 1982; "Draft Report of the Conference," 2nd U.N. Conference on the Exploration and Peaceful Uses of Outer Space, April 20, 1983; *Civilian Space Policy and Applications* (Washington, D.C., Congress, OTA, 1982), Chapter 7.
8. "Joint National Paper," 2nd U.N. Conference on the Exploration and Peaceful Uses of Outer Space, August 17, 1981.
9. Neil Davis, *Japan Times*, August 8, 1982; *Oriental Economist*, August 1982; *Space World*, August 1982; "Japan First with DBS," *Space Flight*, 1983.
10. *Aviation Week and Space Technology*, June 15, 1981, August 10, 1981, October 20, 1981; *Space Age Review*, July–September 1980, September 1981, February 1982; *Satellite Communications*, August 1981; *Christian Science Monitor*, March 9, 1982, September 16, 1982.
11. *Satellite Communication*, November 1981.
12. *Ibid.*, October 1981, April 1982.
13. *Ibid.*, August 1981, November 1981, May 1982, June 1982, September 1982, February 1983, June 1983.
14. *Ibid.*, August 1981, September 1982; *Aviation Week and Space Technology*, April 20, 1981.

15. *DBS News*, November 1983.
16. *Ibid.*
17. *Satellite Communications*, April 1982, May 1982, June 1982, July 1982. Presentations by lawyers Martin Rothblatt and Delbert D. Smith at the L-5 Space Development Conference in Houston, Texas, 1982.
18. *Aviation Week and Space Technology*, November 7, 1983.
19. *DBS News*, November 1983.
20. *Satellite Communications*, November 1981.
21. *Space Calendar*, September 19, 1983.
22. *Space Calendar*, October 3, 1983.
23. *Satellite Communications*, February 1983.

6 REMOTE SENSING

Communications and remote sensing satellites developed in the same period and employed much the same technology. But although communications satellites became profitable by the late 1960s U.S. government remote sensing has yet to attain profitability. Only now is a private remote sensing capacity developing that might attain profitability, although nowhere near the aggregate revenue levels of the communications industry. This is the case because of the structure of the market as well as its size.

THE TECHNIQUES

Remote sensing means simply that a satellite or airplane is measuring or photographing the earth from a remote point. This measurement at a distance may be passive sensing, such as photographing the earth; or active sensing, which involves bouncing light waves off the ground and measuring the spectral frequencies not absorbed by the objects on the earth. Each object has its own spectral signature, which is recorded by the satellite receivers. The data are recorded on computer and can be purchased as tapes or photographs.

The sophisticated satellites today take readings of a location using different instruments as well as measurements in different frequencies. These data are encoded on computer tapes. The real art and the real business in remote sensing have until now been the interpretations of these multiple measures to nurse and tease information from the tapes. An entire industry of value-added, computer interpretation of these tapes has developed to help private companies and governments to assess the data.

As early as 1960 the U.S. Weather Bureau and the Department of Defense cooperated on the first Television and Infrared Satellite (TIROS). Its data, the first ever worldwide cloud-cover pictures, permitted better observation, analysis, and forecasting of weather than ever before. Since 1966 the entire earth has been photographed daily by U.S. satellites.

The TIROS, NOAA, and (until 1978) Nimbus series of satellites provide the service in their 500-mile-high polar orbit. More recently NASA has begun its new series of geostationary satellite (GEOS), which can monitor the entire hemisphere continuously. The instruments carried by GEOS permit it to measure atmospheric pressure and water vapor as well.

The other aspect of remote sensing, the observation of the earth and its geological formations received much publicity when the early astronauts marveled at how much they could see and photograph from orbit. Based on these observations in the early 1970s, NASA designed its Earth Resources Technology Satellite (ERTS), later renamed *Landsat*.

Landsat 1 was launched in 1972 and operated for six years, Landsat 2 operated from 1975 to 1978 and Landsat 3 since 1978. Since 1978, however, Landsat has been plagued by mechanical problems. Landsat 3 malfunctioned and has produced less detailed data than desired. Landsat 4 in 1982 had a series of malfunctions that forced NASA to prepare the satellite for retrieval and repair by the Shuttle. Because of this, NASA had to speed up the launch of the last Landsat, D-prime, for 1984 in order to fill the data gap.

The last two Landsats have possessed new instruments that increase their capacities and capabilities. The Thematic Mapper provides data in seven spectral bands; moreover, the improved multispectral scanners can produce 800 scans per day, compared with the 190 on Landsat 3.

The Landsats inhabit a 570-mile-high polar orbit from which they can cover the entire earth every eighteen days. The data are transmitted from the satellite to earth-based receiving stations (antenna).

The United States collects, stores, and sells the data from the Earth Resource Observation Systems EROS Data Center in Sioux Falls, South Dakota. A dozen other earth stations have been constructed and licensed at $200,000 a year to foreign nations to allow direct reception of Landsat data.[1]

THE APPLICATIONS AND MARKETS

As suggested earlier in the book, remote sensing has not developed as a business, in part because of the size and nature of the market. Unlike communications, a large component of remote sensing applications are public goods.

In the literature of economics, a public or collective good is one in which all members of a group or society share. Neither is the good generally divisable, nor can a consumer be excluded from its use. Collective goods evoke the "free rider" problem. Because of nonexcludability, consumers do not pay their share for consumption and enjoy their use freely or cheaply at the expense of other consumers. Because the product—the data—is a fixed entity, the marginal cost of producing more tapes of the data is minimal. Pricing a public good becomes very difficult because these two attributes, nonexcludability and nondivisability, prevent reliance on supply and demand pricing mechanisms. At the other extreme are pure private goods. The consumers of the good pay for what they want at the price they will accept; supply and demand determines the price.

In a more applied sense, two other issues can be isolated in determining public and private goods here: Is the public good sufficient for all purposes, or does a group of consumers need a unique product? Similarly, how much are those consumers willing to pay for this unique product? These statements define the limits of free ridership; these statements are important in analyzing the applications of remote sensing and whether a market exists for the product as a private good.[2]

Table 6-1 analyzes selected services provided from remote sensing data. The table is organized in terms of the services and nature as public, private or mixed goods. The *mixed* category identifies services that are presently a public good, but many free riders may respond to a unique product if the price were right in the furtherance of their business, converting the good into a private one. The degree

Table 6-1. The Remote Sensing Market—Public/Private Goods.

Selected Services	Public	Mixed	Private
Water quality monitoring	+		
Air pollution monitoring	+		
Earthquake prediction	+		
Weather prediction	+	+	
Fish census	+	+	
Ocean currents	+	+	
Topographical mapping	+	+	
Crop yield	+	+	
Geological structures mapping	+	+	+
Search for water	+	+	+
Forest census	+	+	+

of public or private in this table, however, are qualitative judgments, as will become clear in the discussion to follow.

The first three uses of remote sensing data are included in the pure public goods category. The data have been used to measure air and water quality and to help identify sources of pollution. Yet even here a private sector could be identified—the parties to a law suit involving liability for pollution.

Even as pure a public good as earthquake prediction can be used by a party in a lawsuit to determine whether proper precautions were taken in location and erection of some structure. Such data, however, represent a small, erratic market, and the users could probably rely on the existing data (that is, do not need a unique product). The three services, thus, while important to society, could not generate a private industry in remote sensing.

The next five services in the table begin to offer the foundation for some segmentation of the market, although these too are predominantly public in nature. Billions of dollars and hundreds of lives are saved each year by accurate weather forecasting by satellite. The other services in this category likewise have a large component as public goods. Unlike the first category, these services also are important to a specific sector of the private economy. Ski resorts and other outdoor businesses may be very interested in a localized forecast. Mapping, also basically a collective good, could admit of private uses. Knowledge of ocean and wind currents can cut fuel expenses for ships and airplanes. Remote sensing for fish census can be very important to fishing boats. Moreover, large farming conglomerates would be benefitted in their decisions about planting, irrigation, and harvesting based on crop yields estimated from satellite data.

In these services the free riders are more numerous and also represent a larger potential market (if the price were right, the product useful, and the public goods not specific enough). The private remote sensing products, however, were likely to be too expensive and too few for them alone to justify a private remote sensing industry.

The last set of three remote sensing applications, however, are likely candidates to support an independent product and industry. Large lumber concerns like St. Regis may be very interested in forest surveys so that it can have long-term national planning of the health and the yield of its crop of timber. The service of remote sensing most like a private good is definitely the interest of mining companies in the location of geological formations likely to hold reserves of oil, coal, or other mineral resources. For instance, Landsat data has found that the geological fault associated with Alaska's North Slope oil continues into the Arctic Ocean. The U.S. Space Shuttle also found a potentially rich site of mineral resources in interior Mexico. Many mining and drilling companies increasingly are relying on remote sensing data to decide whether to bid on sites in land auctions or to conduct mining operations on existing sites.

Despite this group of uses, most Landsat remote sensing data have been utilized as a public good in the United States and in the rest of the world. More than 30 American states use Landsat data. California's Department of Water Resources has used the data to identify crops in need of irrigation. The Department of Environmental Resources in Pennsylvania has employed the data to try to limit the damage to forests caused by the gypsy moth caterpillars. New Jersey has surveyed residential developments with this data. Coastal states have used it to battle ocean erosion. The list of U.S. national and state government uses of this data is almost limitless.[3]

The data have also been sought by nations around the world. Twelve nations have paid the U.S. government to construct the earth stations to receive Landsat data directly. These stations are located in Brazil, Italy, South Africa, Sweden, Japan, Australia, Argentina, India, China, Thailand, Mexico, and two sites in Canada.

Canada was an early user of remote sensing. Australia has used the data for crop irrigation and to fight grass fires. The data have also

Table 6-2. Landsat Users, by Percentage.

	1977	1978	1979	1980	1981	1982
Dollar total (in millions)	$1.45	$1.98	$2.13	$2.39	$2.50	$2.94
	\multicolumn{6}{c}{User (%)}					
Federal government	26	31	23	16	19	20
State/local government	1	1	1	3	4	5
Academic	10	8	11	9	8	7
Industrial	28	24	24	26	31	31
Non-U.S.	30	32	36	42	33	33

Source: Based on Table 3, p. 40 of "Commercialization of Land and Weather Satellites," Report of CRS, Committee on Science and Technology, U.S. House of Representatives, 98th Cong., 1st sess., June 1983; "Production Statistics, FY 1982," EROS Data Production Branch, November 1982.

been used to identify the cut and burn of agriculture in Northern Africa as one of the reasons for the southern drift of the Sahara.

Latin American countries have used these data extensively. Brazil is one of the biggest users of Landsat data. Brazil remapped the entire country and found errors measured in the tens of miles in their aerial maps. The Brazilians used the data to draw the route of their new trans-Amazon highway. Peru has used the tapes to study the growth of its deserts. Remote sensing has been important in the search for mineral resources in the difficult, nearly inaccessible terrain of the Andes and the Amazon. During the late 1970s, the Inter-American Development Bank loaned over $1 million a year to Latin American countries for remote sensing projects.

Table 6-2 shows the small but steadily increasing size of the market in Landsat market. Sales doubled in five years from 1977 to 1982. Foreign countries account for one-third of the sales. Note also that less than one-fourth of the data is bought by U.S. government and that over one-third is purchased by private users.

Chevron Overseas Petroleum has been using Landsat data since 1972 and has pioneered many new methods for interpreting the data. NASA also has been cooperating with the Geosat Committee. The Committee presents 100 companies from the oil, ore, and engineering fields. Given the proprietory nature of such geological information, the degree of cooperation is unusual.[4]

By 1982 the trend of economic studies was reflecting an optimistic market for remote sensing data. Table 6-3 supplies the results of these studies. Although the results show a wide discrepancy, they suggest a viable industry at some point in the future.

The Business of Landsat

Through the 1970s, NASA operated a small, financially losing business in Landsat data. In the late 1970s, the House Committee on Science and Technology conducted hearings on the future of Landsat.[5] The consensus of the meetings was that a privately owned Landsat could not become economically viable before 1989. Moreover, only the most optimistic projections called for a breakeven point by the year 2000. The market was assessed as $140 million, but expenses would run between $140 and $400 million. President Carter in late 1979 issued an order to devolve Landsat operations from NASA to The National Oceanographic and Atmospheric Administration (NOAA). It was still hoped that by 1989 NOAA would further devolve the service to private industry.

In 1980-81, NOAA's first year of Landsat operations, NOAA received $6 million in revenues but expended $30 million to operate the program. In 1982 NOAA raised its fees dramatically: the cost of 70 millimeter black-and-white film rose from $8 to $26 and the digital tapes from $300 to $600. Although the price rise did redress the balance somewhat, NOAA's decision is even more important in its implications for the privatization of remote sensing—as an attempt to ween some users away from the government subsidies (as a public good).

Figure 6-1 shows that the quantity purchased generally goes down after each price rise, but the amount of data purchased does rebound. This phenomenon suggests that, despite the resistance, the buyers return for the data that they need.

In 1982 President Reagan announced that he would try to transfer Landsat to private enterprise by 1984, five years ahead of schedule. He included the weather satellites in this plan. The decision was controversial both in Congress and in the states.

The states had become used to cheap prices and extensive technical help and feared that both would be missing in the privatized Landsat operation. The National Conference of State Legislatures

Table 6-3. History of Remote Sensing Market Projections.

Author	Date Published	U.S.	Worldwide	Remarks
Earth Satellite Corporation Booz, Allen, and Hamilton	1974	0.13 (1985)		Landsat 1, resource surveillance only
Econ	1974	(1975)	3.3–6.3	Landsat 1, broad application
ABT Associates	1981	5–10 (1982)	30–35	New sensors, aggressive marketing, total benefits
OAO Corporation	1981	3–5 (1982)		Potential market for data
Terra-Mar Associates	1982	0.35 (1985) 1.54 (1990) 12.6 (1995)	0.55 3.82 21.0	Realizable market for data

Source: TRW, "Space Station Needs, Attitudes and Architectural Options Study," for NASA, April 1983.

REMOTE SENSING 83

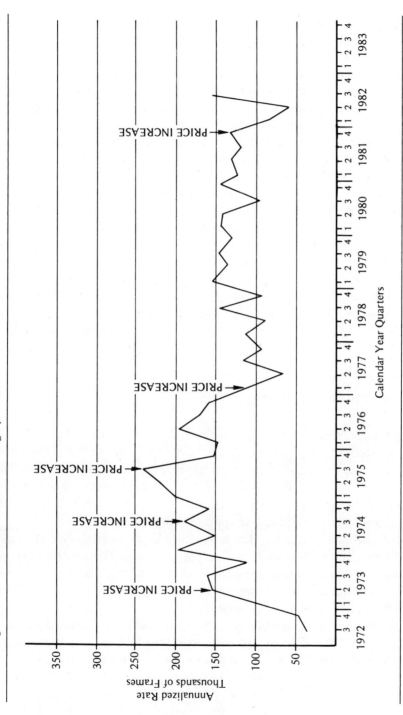

Figure 6-1. EDC Sale of Landsat Imagery Frames.

Source: "Production Statistics," EROS Data Center, November 23, 1982.

and the National Governors Association complained about both the Carter and Reagan plans for the denationalization.[6] Congress has since put the White House on notice that it will review any decision to sell the remote sensing and weather satellites to private industry.

The Department of Commerce has, however, continued with its request for proposals (RFP) for private companies to purchase these satellites. Several companies have considered such a purchase. Comsat General showed early interest in operating the weather as well as the earth resources satellites. The chairperson of Comsat, John Johnson, testified that Comsat was uniquely qualified to operate all of these NOAA and NASA programs. At least five other companies have expressed interest in Landsat; they are RCA, General Electric Space Systems, Lockheed, Bendix, and Computer Science Corporation.

The technology of the middle 1980s, however, suggests that Landsat may no longer be the way to proceed for private enterprise in remote sensing. Moreover, the government's price increases to make Landsat more viable as a private company may have backfired. The higher prices have given the new private companies a more attractive margin in which to operate and to attempt to underbid the Landsat data.

The Private Options

Hughes and Comsat, two space giants, have both studied private satellite alternatives to Landsat. The U.S. Space Shuttle has widened the alternatives available for private companies in remote sensing. The Itek Corporation considered flying a large-format camera on the Shuttle but has been hesitant because of the unclear situation with the government-subsidized (Landsat) competition.

A California company, Terra Mar, has incorporated with the goal to launch its own satellite for remote sensing. The profitability of this and other companies will depend on the interest of the oil and mineral companies in the potential and price of these new satellites in finding new deposits of resources.

Space America, another space venture, will also depend on this market. In 1982 SSIA joined with American Science and Technology Corporation and Aeros Data Corporation in the Space America venture. The group petitioned the FCC for clearance to orbit three satellites. The first satellite to be launched in 1986 will have four thematic mappers (three at 80 meters; one at 43 meters), two Coastal Zone Color Scanners (80 meters), and a multispectral scanner (43 meters). The later satellites will have improved equipment (20 meters); all will fly in a 500 mile sun synchronous orbit (meaning the craft is always over the same spot of the earth at the same time of day). Space America with Bendix Corporation also has bid on the privatization of Landsat.[7]

FOREIGN REMOTE SENSING ENDEAVORS

Remote sensing has important military as well as civilian applications throughout the world. The Soviet Union, Japan, France, India, and other countries have developed series of remove sensing satellites. Following up on its successful pattern in Arianespace, France established the company Spot Image to market the data produced by the new Systeme Probatoire d'Observation de la Terre (SPOT) satellite due for launch in 1985. The satellite will have a 20 meter (65.6 foot) resolution in its multispectral sensors and a 10 meter (32.8 foot) resolution with black and white.[8] The French Space Agency (CNES) owns 39 percent of the shares in the new concern. Other French investors make up a majority of the company, but other nations are also represented.

The company has established an office in the United States and in several other countries. It has conducted an aggressive marketing campaign. Unlike the rigid Landsat fees, Spot Image is setting its fees to correspond to use, and its earth stations will be compatible with both Landsat and SPOT satellites. India became the first third-world country to sign a contract with Spot Image. The earth stations will cost $1.5 million.[9]

In addition to the French SPOT, ESA will launch its Earth Remote Sensing satellite (ERS-1) in 1988. Japan will follow with its own ERS-1 in 1990. Even sooner, India plans to launch its Indian Remote Sensing (IRS) satellite on its own launcher. The Indians will rely on a Soviet launcher to orbit their second IRS in 1986.[10]

The most recent entry into the multinational drive to enter a nascent market in remote sensing is Sparx Corporation. Its history is tied to the development of the U.S. Space Shuttle. On STS-7, the U.S. Shuttle flew a German payload, the SPAS-01. Messerchmitt-Boelkow-Bohm (MBB) developed SPAS and its modular optoelectronic multispectral scanner (MOMS). This scanner has a 65-foot resolution. The SPAS will be very flexible, at first remaining attached to the Shuttle, but later to be free-flying and retrieved by the Shuttle. Sparx envisions a fleet of five to ten Shuttle Pallet Satellites by the end of the decade.

Sparx is a joint venture of two groups—MBB and the Stenbeck Group, the New York based Swedish investment company. (Originally it was a tripartite agreement; Comsat has since bowed out.) The group plans to invest $50 to $75 million by 1986. MBB will provide equipment and Stenbeck, financial backing.[11]

This enterprise, like the others, has targeted the mineral and oil interests as the prime market, although President Klaus Heiss (formerly of Space Transportation Company) expressed the opinion that agricultural information may become an important market in time. Since the SPAS pallets cost only 10 percent of a Shuttle mission, the mission costs will not be excessive. Unlike Landsat data, the SPAS data will be proprietary—a pure private good.

Many countries have always objected to remote sensing, especially in its proprietary mode, as a violation of their sovereignty. Since Eisenhower the U.S. position has always been an open sky policy. As long as Landsat data were open to all, this issue had never been directly challenged. Sparx—a multinational consortium—may provide its first real test.[12]

THE POLITICS AND ECONOMICS OF REMOTE SENSING

The market in remote sensing will remain competitive and speculative; Table 6-4 demonstrates that types of data may present users with many choices. The U.S. government has sought to bring legal order to the confusion in the field. Although the United States has been unable to negotiate an international treaty to legitimize private remote sensing across borders, the process appears to be developing as a fait accompli. Eisenhower's open skies policy will soon be tested against the sovereignty of other countries. Domestic and international courts may have to decide whether the sensing of foreign ter-

Table 6-4. Planned Sensors Compared.

Satellite	MSS (meters)	TMT (meters)
Landsat	80	30
Space America	80	43 (later 20)
SPOT	20 (10 black and white)	—
SPAS	20	—

rain, with private proprietary interest in the data, violates the sovereignty of the sensed nation in the form of some kind of invasion of privacy. U.S. courts have, indeed, found an invasion of privacy and a violation of the Fourth Amendment of the U.S. Constitution when the Environmental Protection Agency photographed a U.S. industrial complex from the air without a warrant or permission.[13]

The government needs to continue to develop needed new technologies and to conduct services that are initially unprofitable or are necessary for security. But operational high technology is best conducted by private industry—in part because of the superiority of private industry in business and in part because of the need of government to be free to develop the cutting edge of technology. This division of labor has been the most efficient strategy for the United States and promises to continue to be.

The United States also will need to balance the needs of the Landsat program against the growth of the private ventures. The situation is like that of space vehicles several years before—a government program as an expensive proof of project competing against a growing foreign market and at the same time nurturing domestic alternatives. Legislation has been proposed that is intended to accomplish these diverse purposes (see Chapter 9).

NOTES TO CHAPTER 6

1. *Civilian Space Policy and Applications*, Washington, D.C.: Office of Technology Assessment, U.S. Congress, 1982, pp. 124 *et seq.*; "Landsat," NASA Pamphlet, 1981; *Aviation Week and Space Technology*, September 7, 1980, October 27, 1980, September 15, 1981; Robert Hirsch and Joseph J. Trento, *The National Aeronautics and Space Administration* (New York: Praeger, 1973), pp. 26-36.

2. Norman Frohlich and Joe A. Oppenheimer, *Modern Political Economy* (Englewood Cliffs, N.H.: Prentice-Hall, 1978), pp. 32 *et seq.*
3. *Space Age Review*, October/November 1980; *Aviation Week and Space Technology*, July 17, 1978, October 27, 1980; *Satellite Communications*, February 1982; *NASA Activities*, October 1981.
4. *Aviation Week and Space Technology*, June 22, 1981, November 7, 1983.
5. *Ibid.*, April 20, 1981.
6. *Ibid.*, July 14, 1980, April 20, 1981, July 27, 1981; *Space World*, April 1983.
7. *Space Calendar*, October 3-9, 1983, October 24-30, 1983, November 21-27, 1983.
8. *Aviation Week and Space Technology*, January 1, 1982, March 14, 1983.
9. *Space Calendar*, November 7-13, 1983.
10. *Ibid.*
11. *Aviation Week and Space Technology*, November 7, 1983; *The Commercial Space Report*, November 1983.
12. *Aviation Week and Space Technology*, November 7, 1983.
13. *Dow Chemical v. U.S.*, 536 F. Supp. 1355, E. D. Michigan, 1982.

7 MANUFACTURING, MINING, AND ENERGY

Space manufacturing, as esoteric as it sounds, will be a reality in the late 1980s. NASA has made a concerted effort to interest private enterprise in using the U.S. Space Shuttle to develop this entirely new industry in space. Between 1976 and 1979 at least five major studies were conducted on the nature of the coming Shuttle era, by the American Institute of Aeronautics and Astronautics, General Electric, the National Academy of Sciences, Rockwell International, and Science Applications. The Rockwell study, in particular, found that American companies in the 1970s knew very little about the potential of the Shuttle but were ready to enter any space competition their rivals entered. Profitability and risk were the big question marks in all studies.[1]

Undaunted, NASA in 1978 created its Material Processing Space Division and other activities directed toward future space commerce. The Shuttle did reduce the costs of doing business in space in at least two ways. The Shuttle is more versatile than earlier space transportation; its cargos can range in size from a 200 pound package to a 60,000 pound space station. Moreover, the human element should not be overlooked: Missions have been saved many times by the dexterity and ingenuity of the astronauts. The Shuttle thus provides one of the most basic needs of permanent and profitable manufacturing in space.

The Shuttle may never be as economically successful as planners had originally hoped, however. As a result, NASA has resorted to several smart marketing ploys. One technique has been the Getaway Special (GAS). NASA, thereby, agreed to launch small packages for scientific or commercial applications. These packages had to be under 5 cubic feet and 200 pounds. The company, university or other organization pays $500 earnest money and $10,000 total payment.

The program was a tremendous success. In the first year the number of Getaway Specials reached over 100 and by 1983 had risen to more than 300. Among the GAS holders are Battelle Memorial Institute, Columbia Pictures, Corning Ware, Coors Brewing Company, Dow Chemical, Dupont, and Ford Motor Company. Many foreign companies and nations also took advantage of the offer. ESA, Egypt, Japan, and the German Volkswagen all purchased getaway specials. But the biggest customer of the GAS has been the West German government, reserving fifteen specials on the Shuttle.[2]

The GAS has been a success both in its own right and as part of a greater policy. GAS opened up the Shuttle to new users in the schools and the industry. It creates friends, constituents, and users of the space agency. NASA, moreover, has often included the GAS as part of a commercial package in order to lure these new users into space commerce by reducing the cost and the risk of testing new products in space.

NASA also developed two government-business working relationships: the Joint Endeavor Agreement (JEA) and the Technical Exchange Agreement (TEA). In 1981 McDonnell Douglas and Johnson and Johnson signed a JEA with NASA. The agreement commits NASA to grant exclusive use of a specific *continuous flow electrophoresis* (CFE) process for the manufacture of biological products, including pharmaceuticals, to those companies. This space-based electrophoresis takes advantage of the unique attributes of space. On earth, this process requires placing a compound in suspension and charging it electrically. The charged particles in the compound are then collected at the positive- and negative-charged poles. The charged particles tend to be swept back into the solution by convection currents created in the process. In the microgravity of space, however, these currents do not form, and the separation of particles is more complete. Therefore the same process in space can produce both a purer and a larger quantity of the product.

McDonnell Douglas and Johnson and Johnson were allowed to fly free of cost electrophoresis experiments on the Shuttle in 1983. The agreement called for three more free flights in 1984 and two in 1985. The initial experiments proved the process and produced 500 times the quantity that could be produced on earth. In 1986 or 1987 NASA would also launch an 8,000 pound automated factory to produce the still undisclosed pharmaceuticals. NASA would also provide the power (3.5 kilowatts) and at least some of the labor. McDonnell Douglas and Johnson and Johnson would conduct experiments for NASA as well and reimburse NASA for later flights to the factory for its supply, repair, and retrieval.[3]

Other types of electrophoresis have received attention:

1. Grumman has identified *large pore cell electrophoresis* as a $200 million market for producing pure enzymes, isoenzymes, and hormones for diagnostic kits.

2. Rockwell finds that interferon production could be doubled using *recycling isoelectric focusing* (RIEF). In this process, charged protein ions migrate to the points associated with the pH factor (acid to base). RIEF is somewhat superior to CFE because it is more independent of gravity disturbances.

3. Finally, products can be separated according to their densities, much like oil and water. In space, there would be little gravity to cause sedimentation.

Microgravity Research Associates (MRA), a Florida firm, has concluded a JEA with NASA that includes seven free flights aboard the Shuttle. The flights will allow MRA to perfect the electroepitaxy process to grow crystals. The plant, a 30,000 pound factory on the Shuttle, would grow larger, purer crystals of gallium arsenide, a semiconductor for the lucrative computer industry. By the late 1980s or 1990s, the present silicon chips will be approaching the theoretical limits of their capacity to carry data (through the dopants or impurities in the crystal). Gallium arsenide should be the new product, with ten times the capacity of silicon, waiting to take its place as the more efficient semiconductor.[4]

Other companies have publicized plans to offer space platforms for hire (to conduct manufacturing experiments in space). Fairchild Space and Electronics Company has advanced plans for the Leasecraft satellite. In 1983 NASA and Fairchild signed a JEA. NASA will launch the craft in 1987 and retrieve it a few months later. It should have a lifetime of more than ten years. These first two flights (normally costing $50 million) will be free of charge; Fairchild and its clients, however, will reimburse NASA for later flights. Fairchild will spend $200 million to develop the automated factory and is negotiating with McDonnell Douglas and Johnson and Johnson to fly the continuous flow electrophoresis system.[5]

The two foreign endeavors must be mentioned in this regard: the German SPAS can be used for manufacturing as well as remote sensing. Eureca, the new ESA-designed shuttle retrieval satellite, will also be able to conduct scientific experiments. Other American companies are also hoping to get into the business of providing facilities and services to other space manufacturers that presently have the experience in production but not in space. Ball Aerospace Systems plans to have a material processing facility by 1986 and would lease sections of the craft to private commercial activity.

At least two efforts, by the Foundation for Space and by Space Industries, Inc. (SII), have looked seriously into human-tended, private stations to carry out space research and production. Furthest along is SII, which has signed a memorandum of understanding with NASA. Headed by Max Faget, former director of engineering development at NASA in Houston, the company has advised NASA (including its own independent assessment of the eight studies by aerospace companies in 1983) of the specifications and benefits of an inhabited space station. The company probably would begin with a craft similar to Fairchild's or Ball's, but its ultimate goal is a private space station or a private module attached to a U.S. space station.[6]

In addition to the basically private efforts, NASA also would like to have a space station by the early 1990s for science, for servicing of satellites, for a launching pad to geosynchronous orbit or the planets and for commercial activity in space. It was for this purpose that NASA let eight $800,000 contracts to major aerospace firms in 1983 to investigate the uses for space stations. The studies suggested that space stations could indeed provide the infrastructure for a multibillion dollar industry in space manufacturing.[7]

In his 1984 state of the union address, President Reagan dramatically directed NASA to orbit a space station within a decade. The announcement climaxed a more than three-year effort by space advocates to convince the White House of the necessity of a permanent human presence in space.

Private industry also recognizes the importance of an astronaut-tended space station. The manager of McDonnell Douglas's project with Ortho Pharmaceuticals, James Rose, estimates that the untended option (such as Leasecraft) would result in three space-based products on the market in the 1990s. With a permanent space station, however, the manufacturing costs could be reduced almost a quarter. The ongoing research and the cost savings should result in maybe fifteen products reaching the earth market in the 1990s.[8]

The studies identified semiconductors, glasses, optical fibers, pure crystals, alloys, and pharmaceuticals that could be profitable taking advantage of unique attributes of the space environment. Boeing Company, for instance, cited a market for gallium arsenide of $170 million in 1990, increasing to $5 billion by the year 2005. McDonnell Douglas identified twelve promising products, including interferon and beta cells.[9] Table 7-1 lists these twelve products as well as their use, status, and potential market.

Other companies have begun to conduct research on whether space may be a cost-effective site for producing old or even entirely new products. The formula for space profitability will be a product that is small (because space transportation is expensive) and takes advantage of some unique attribute of space to increase the value sufficiently over the same product produced on earth. These attributes include micro (near-zero) gravity, near absolute zero temperature or extremely high temperatures depending on the positioning of the spacecraft. A final advantage of space is its isolation from the earth. This attribute could result in dangerous activities, such as recombinant DNA research, being conducted in space in the 1990s.

Table 7-2 demonstrates that there is a lot of interest generated by NASA's bargain basement pricing strategies. The takers have included both established Fortune 500 firms and new endeavors made up of former NASA engineers or space enthusiasts. Obviously many of these efforts will fail; some (not on the list) already have bowed out after their initial studies. But many of these processes look promising and profitable.

Moreover, technological advances could destroy presently attractive space markets. Biogenetic products may challenge space manufacturing as the best way to produce these substances. This competition would be reminiscent of the developing competition between satellite communications and earth-based optical fiber cables.

Space manufacturing is posed to begin. By the late 1980s, automated space factories will have begun to produce their first output. The 1990s will see the expansion and maturation of this industry. Systems will become more complex and will likely need to be

Table 7-1. Twelve Typical Candidate Pharmaceutical Products for Space Manufacture.

Typical Products	Beneficial Medical Application	Function/Status	Annual Patients (U.S.)
α_1 Antitrypsin	Emphysema	Only research quantities now	100,000
Antihemophilic factors VIII and IX	Hemophilia	100% terminal by age 40	20,000
Beta Cells	Diabetes	Possible single-dose cure	600,000
Epidermal growth factors	Burns	Replacement skin grafting	150,000
Erythropoietin	Anemia	Replacement transplants or transfusions	1,600,000
Immune Serum	Viral infections	EOS provides higher purity	185,000
Interferon	Viral infections	Potential may be unlimited	> 10,000,000
Granulocyte stimulating factor	Wounds	Only research quantities now	2,000,000
Lymphocytes	Antibody production	Replace antibiotics chemotherapy	600,000
Pituitary cells	Dwarfism	Currently not curable	850,000
Transfer factor	Leprosy; multiple sclerosis	Potential for other applications	550,000
Unokinase	Blood clots	Low development costs	1,000,000

Source: McDonnell Douglas, Space Station Report for NASA, 1983.

human-tended. Increased power will be required probably from solar cells. Before the century is out, the logic of this development will lead to a modular, industrial park in space. NASA or an international Western consortium will provide most of the supplies, power, and maintenance modules and the people; individual modules will be operated by private U.S. businesses as well as by foreign companies and nations.

ENERGY AND MINING

Sometime early in the next century our space-based commerce will move beyond low earth orbit. Manufacturing and remote sensing will generally be limited to lower orbits. Already our communications satellites populate the geosynchronous orbits. These orbits are becoming congested, and technology is seeking methods to alleviate the crowding so that most of the seemingly endless communications services can be provided for the earth.

One of these new technologies is the large space antenna or antenna farms—a large space platform with a large solar array and equipment to handle a multitude of services. This multipurpose satellite will be able to take over the duties of several, reducing the congestion in orbit substantially. These structures, however, will need to be constructed and maintained. The orbital transfer vehicles (OTVs) of the future will need to be larger, and a new species of OTV will need to be habitable to ferry workers to construct and maintain the large space structures. If the radiation problems can be overcome, nearly permanent crews will be kept in geosynchronous orbit to service the dozens of large space structures of the twenty-first century.

Another use of geosynchronous orbit that has received a lot of consideration is the solar power satellite (SPS). This satellite would definitely require many workers and maintenance personnel in space. Although NASA has long used solar energy for many satellites, it was Peter Glaser, presently a vice-president of Arthur D. Little, who in 1969 published the idea of space-based solar energy to power homes and industries on earth.

These satellites could stretch for miles with large arrays of photovoltaic solar cells. Capturing the solar energy, the satellite would then convert it into microwave energy and beam the energy down to earth receivers (rectennas). These receivers would occupy acres of land;

Table 7-2. Space Manufacturing—Companies and Proposed Projects.

Companies	Proposed Products
Glass	
Westinghouse Corporation	Glasses for lasers and telescopes
Oakridge National Laboratories/ Union Carbide	Glass-forming alloys
Semiconductor/Crystals	
Westech Systems	Container for space research
Microgravity Research Associates	Gallium arsenide
Johnson Mathey (United Kingdom)	Iridium crucible
Honeywell Corporation	Crystal growth
Biologicals	
Lovelace Medical Foundation	Cell growth
Ortho Pharmaceuticals	Cell separation
Battelle Memorial Laboratories	Collagen fibers
Metal Processing	
Aluminum Company of America	Aluminum refining
International Nickel Company	Electroplating metals
John Deere/Bethlehem Steel	Graphite in cast iron
Anonymous company	Coal to coke
Others	
Dupont	Catalysts in space
Calcitet	Composite materials for bone substitutes
International Space Corporation	Unstated product
3M Corporation	Unstated product

Source: Compiled from *Space Calendar*, October 24–30, 1983 and later issues.

even so, such a system would provide the energy needs of entire cities.

The dream of cheap and available worldwide energy is alluring; the reality is a little harder to assess. The balance sheet of the SPS looks good against earth-based powers of the twenty-first century; however, the balance sheet requires several assumptions that might not be met. We might find more oil and gas than expected; we might have a breakthrough in shale oil or other synthetic fuel processes. Problems in nuclear energy, especially fusion, might be solved. Conservation may prove even more effective. On the other hand, SPS may prove to be more expensive than planned.

A list of potential problems hints at some of these unexpected costs. We have almost no experience with construction in space; unanticipated problems could greatly increase the cost. The effects of microwave energy on the atmosphere or on birds or planes flying through the beam is not clearly determined. Similarly, some critics have asked what the effects on people and property would be if the energy beam strayed from the receiver and cut a swath through the countryside.

At the end of the 1970s, the U.S. Department of Energy (DOE) created an SPS Project Office to study the microwave SPS plans. The office involved three interest groups in its work. The L-5 Society represented the pro-space views; the Citizen Energy Project stood for the environmentalists and Forum for the Advancement of Students in Science and Technology (FASST), a now defunct student science organization, represented general scientific attitudes.[10]

The DOE study complemented studies by the National Academy of Sciences and the Congress's Office of Technology Assessment. These studies found that many of the fears were simply unfounded: The power beam could not stray and destroy property, because it would defocus if not controlled by its earth-based pilot beam. More work needed to be done on potential harm caused by microwave radiation, but the earlier studies were encouraging. The size of the satellite would protect it from all but a huge meteor. Moreover, much of the construction would be accomplished by robotic devices, with which the Shuttle crews are already gaining experience.

Although DOE estimated the project to cost over $3 trillion for sixty satellites by the year 2030, it found no show stoppers. The OTA and the NAS studies found the idea promising but premature.[11]

SPS also faces problems in the domestic political arena. Environmentalists were activated by these studies to oppose SPS. Other analysts attack the program as "centralized technology" and explicitly or implicitly laud "decentralized" technology as important for freedom. Larry Luton has offered an argument, however, that defuses even this contention. He points out that the SPS work will also facilitate the earth-based solar energy with improved photovoltaic cells. Sunny, equatorial climes thus will be served well by earth solar power (decentralized). More polar latitudes, however, will be served better by space-based solar energy.[12]

Other SPS options also counteract some of the negative reaction to its initial reception. Laser transmission and reflected sun light (by space mirror) are alternatives that may be less complex, less costly, and less objectionable to environmentalists. Also, the cost estimates on these SPSs have always been based on structures composed of materials from earth, manufactured on earth, and ferried into space. Sometime in the early twenty-first century, nations will begin to mine ores in space and even to refine them there. At some point this process could prove cheaper than bringing the materials from earth. The irony of using space materials is explained by atmospheric drag and by the nature of the gravity well. The force of gravity means that the thrust required just to get into orbit is most of the thrust necessary to go to the moon. It is also easier to fall into the well than to climb out of it. So it will require less fuel to carry building materials from the moon to earth's orbit than to get them from the earth to orbit.

CELESTIAL MINING

The gravity-well phenomenon also has tremendous implications for deep space industrialization and colonization. With or without SPS, we will need to develop orbital transfer vehicles that can take large structures and people out to the geosynchronous orbits. An OVT that can get to geosynchronous orbit can get to the moon. This incidental effect means that we will have the infrastructure to return to the moon in a big way by the end of the century.[13]

During Apollo, the astronauts discovered iron, aluminum, and titanium on the moon. Moreover, oxygen was found to be abundant in the rocks. Oxygen may be the most important "ore" in space: in addition to breathing it, we will need it to combine with hydrogen (rare on the moon) for water and, most important, to use it as a fuel for our spacecraft. The effects of gravity suggest a scenario in which it may prove cheaper to import our oxygen from the moon to earth's orbit rather than from the earth to earth's orbit. The lunar regolith may be used for radiation shielding to protect satellites.

Moon mining may become feasible around the turn of the century, but at first its product will only be used for local consumption. It will take years of improvements in technology to lower the cost substantially. Yet, at some point, as the cost of mining the earth of increasingly scarce resources relentlessly rises, the comparable costs of earth and space ores will cross, and we will begin to import resources from space to the earth itself.

Gerard K. O'Neill, at Princeton University, has spent over a decade developing the scenario and the hardware to mine the moon and asteroids. This scenario envisions electromagnetic mass drivers or catapults on the moon, to hurl the ore into space. The ore would be caught by a "mass-catcher" located behind the moon and then be placed into an orbit for processing. Almost free solar energy will be used to smelt the ore. The ore can then be shipped to earth or to earth orbit for use.

In addition to the moon, the asteroids and the planets are future sources of natural resources. Analysis of meteorites on earth and spectral analysis by telescope suggests that the asteroids are a rich source of precious and valuable minerals. Some forms of asteroids also contain the hydrogen (water), carbon, and other ores that could be necessary for a true cislunar society and economy.

Gravitational effects make planetary mining most unlikely, but asteroid mining may be much simpler. This is especially true of the hundred or more asteroids that have orbits bringing them within a few million miles of earth. It would be relatively easy for a space prospector of the twenty-first century to identify a promising asteroid to mine it in place or strap on rockets or mass drivers to nudge it out of its solar orbit into an orbit around the earth. In orbit, the miners could again use solar energy to process the ore. The slag would be a heat shield that would burn off as the ore was pushed out of orbit for an ocean landing. The ore could be processed with large vacuum bubbles in it to give it a specific gravity less than water so that it could float on water.[14]

CONCLUSIONS

Only communications satellites (and to a lesser extent, launch services) have been established as a viable space industry. Remote sensing is a proven technology but not yet a proven market. The commercial push in remote sensing will come with the adjusting of the technology to meet the needs of the consumers.

Space manufacturing is not yet so well developed and needs more than just a fine tuning. The basic research will be conducted in the rest of this decade. By the 1990s, the general space laboratory will give way to the more industrial space factories, although they will probably share the same power and supply sources as the laboratories. Space mining and energy, beyond the end of this century, must remain the gleam in the eye of the future space entrepreneurs still in grade school today. Between then and now lies a major multinational, multicorporation competition for these space markets.

NOTES TO CHAPTER 7

1. *U.S. Civil Space Programs, 1958-1978*, report prepared for the Subcommittee on Space Science and Applications of the Committee on Science and Technology, U.S. House of Representatives, 97th Congress, 1st Session, 1981, pp. 800 *et seq.*
2. *Aviation/Space*, September 1981; *NASA Space Activities*, 1981, 1982.
3. *Wall Street Journal*, October 1981; *Aerospace* (AIA), Fall 1982.
4. *Space Calendar*, October 24-30, 1983; Jerry Grey, *Beachheads in Space*, (New York: MacMillian Publishing Co., 1983), p. 163.
5. *Space Calendar*, October 3, 1983; *NASA Activities*, October 1983.
6. *Austin American Statesman*, December 12, 1982; *Space World (NSI)*, December 1982; Conversation with Dr. Max Faget and Dr. Larry Bell.
7. The Space Station Studies for NASA.
8. *Space World*, November 1983.
9. *Space Business News*, July 18, 1983; *Space Calendar*, July 18, 1983; Boeing Company, "Space Station Needs, Attributes and Architectural Options Study," 1983.
10. U.S. Department of Energy, Office of Energy Research, January 1980.
11. "Some Questions and Answers Asked on SPS," U.S. Department of Energy: Office of Energy Research, January 1980.
12. Larry Luton, "Satellite Solar Power Systems Providing for a Flexible and Stable Energy Future," in Paul Anaejionu, Nathan C. Goldman, Philip J. Meeks, eds., *Space and Society: Choices and Challenges* (San Diego, CA.: American Astronautical Society, 1984).
13. Wendell Mendell and Michael P. Duke, Lunar and Planetary Institute, 1983.
14. *Science Digest*, October 1978; *U.S. Civil Space Programs 1958-1978*, report prepared for the Subcommittee on Space Science and Applications of the Committee on Science and Technology, U.S. House of Representatives, 97th Congress, 1st Session, 1981, pp. 1029 *et seq.*; Robert Powers, *The Coattails of God* (New York: Warner 1981), pp. 137-149; David R. Criswell, *Extraterrestrial Material Processing and Construction*, NASA Final Report, Pt. 30, 1978. O'Neill's Space Studies Institute publishes a newsletter that updates the progress on his space industrialization efforts. See *L-5 News*, February, July, and August 1983.

SPACE BUSINESS AND SPACE POLICY

National Aeronautics and Space Administration

Lyndon B. Johnson Space Center
Houston, Texas 77058

8 DOMESTIC SPACE BUSINESS

Until the 1980s most space businesses were for and from the government.[1] The same companies that did space did aviation, electronics, or automotive work as well as work in the military and civilian sectors. The space work in general, and civilian space work in particular, has always been a small percentage of the business sales of most established aerospace firms. In 1981 space products contributed nearly 15 percent of aerospace sales, totaling almost $10 billion.[2] Because the space percentage of their business is small, the aerospace firms have rarely been at the cutting edge of space politics or space commerce. They usually provide the technical and monetary support but rarely the creative push.

BUSINESS AS USUAL

Traditionally the government has been the sole buyer of space goods and services. The companies thus were relieved of both the need and the opportunity to be entrepreneurially creative. Yet students of the aerospace industry have noted the dangerous business position in which defense industry, including the aerospace companies, finds itself. Although the contracts for which a defense company bids often run to the billions or tens of billions of dollars, the company

Table 8-1. Aerospace Industry Sales by Customer, Calendar Years 1967–1981 ($ millions).

		Aerospace Products and Services			Nonaerospace[a]	
		U.S. Government				
Year	Total Sales	Defense Department	NASA and Other Agencies	Other Customers[b]	U.S. Government	Other Customers[c]
			Current Dollars			
1967	24,130	12,901	4,219	4,430	1,750	830
1968	25,927	13,609	3,978	5,791	1,568	981
1969	25,278	13,832	3,369	5,378	1,633	1,066
1970	24,924	14,011	3,000	5,269	1,465	1,179
1971	22,064	11,877	2,779	4,885	1,372	1,151
1972	21,512	11,195	2,649	5,022	1,546	1,100
1973	24,744	11,846	2,459	7,096	1,925	1,418
1974	27,145	12,329	2,608	8,141	2,060	2,007
1975	30,356	13,795	2,838	8,931	2,496	2,296
1976	31,528	15,106	2,938	8,173	2,879	2,432
1977	33,854	16,023	3,012	8,715	3,625	2,479
1978	38,939	16,770	3,151	12,205	3,860	2,953
1979	44,210	17,708	3,453	15,334	4,087	3,628
1980	52,896	20,994	4,106	18,977	4,762	4,057
1981	63,211	25,490	4,700	21,137	5,986	4,898
1982	66,958	32,225	4,863	18,341	6,341	5,188

Constant Dollars (1972 = 100)

Year				
1967	30,521	16,318	5,336	5,603	2,214	1,050
1968	31,411	16,488	4,819	7,016	1,900	1,189
1969	29,125	15,937	3,882	6,197	1,882	1,228
1970	27,254	15,321	3,280	5,762	1,602	1,289
1971	22,981	12,371	2,894	5,088	1,429	1,199
1972	21,512	11,195	2,649	5,022	1,546	1,100
1973	23,412	11,208	2,327	6,714	1,821	1,342
1974	23,621	10,728	2,269	7,084	1,793	1,746
1975	24,176	10,987	2,260	7,113	1,988	1,829
1976	23,865	11,434	2,224	6,187	2,179	1,841
1977	24,211	11,459	2,154	6,233	2,592	1,773
1978	25,951	11,176	2,100	8,134	2,572	1,968
1979	27,161	10,879	2,121	9,421	2,511	2,229
1980	29,824	11,837	2,315	10,700	2,685	2,287
1981	32,776	13,368	2,420	11,522	3,006	2,459

a. Products and services other than aircraft, missiles, and space vehicles and parts, produced by establishments whose principal business is the development and/or manufacture of aerospace products.

b. All civil sales of aircraft (domestic and export), commercial space sales and all military aircraft and missile exports, including both commercial (manufacture-to-foreign government) and Foreign Military Sales (FMS)/Military Assistance Programs (MAP).

c. A comprehensive revision of the AIA aerospace industry sales series for 1967–1980 was completed in 1982 in order to incorporate different data sources selected to reflect better the evolving composition of the aerospace industry.

Source: Aerospace Industries Association. Aerospace Facts and Figures, 1982–83 and 1983–84.

also dies by the big contract it fails to get. It is a boom-or-bust business.[3] The government's bailing out Lockheed is an example.

The defense industry has traditionally been characterized as an oligopoly, at least among the prime contracting corporations, with a monopsonistic, or predominant, buyer—the government (the Department of Defense (DOD) or NASA). Notice in Table 8-1 that this truism has become less true over the past decade. Just over 20 percent of the sales were to non-U.S. government customers in 1967. By 1981 over 40 percent of the sales were to customers other than the U.S. government. Even the civilian aviation business has been very cyclical. The aviation industry in the 1970s and early 1980s has been a sick industry with much excess capacity. The space side of the industry has, similarly, been a boom-or-bust business, riding high through the middle 1960s and hitting bottom until the 1980s. Only the communications industry has escaped this cycle thus far.

The aerospace industry has attempted to diversify, entering as many sectors as possible to offset this cyclical trap. Companies have absorbed other firms in nonaerospace fields. Another trend has been for the large company to integrate vertically, buying out the companies to whom they used to subcontract business. Between 1968 and 1975 the number of subcontractors declined from 6,000 to 4,000. Many went bankrupt, but many subcontractors were bought out.[4]

The reality of the aerospace (and other defense) industry remains the dependence of the government on the industry and the dependence of the industry on the government. Jacques Gansler, a student of the industry, rejects the folk wisdom that DOD and NASA rotate the prime contract to keep the small number of suppliers in business. He argues instead that the company losing one or two major contracts is so close to bankruptcy, it is forced to make a life-or-death bid to get the next contract.

Offsetting this dire situation somewhat is the subcontracting procedure prevalent in the defense industry, including the space contractors. On average, 40 to 70 percent of the work on a major contract will be let out to other companies—to specialty houses with a limited range of expertise and to major competitors. For instance, Rockwell International had the prime contract for the U.S. Space Shuttle. Other participants in the Shuttle include Boeing, General Dynamics, Hughes Aircraft, Martin Marietta, McDonnell Douglas, Northrop, TRW, and United Technologies. (Table 8-2 shows subcontractors built which parts of the Shuttle.)

Table 8-2. Shuttle Contractors.

Company	Product
Aerojet Liquid Rocket Company Sacramento, California	Orbiter maneuvering system engines
AiResearch Manufacturing Company of California Torrance, California	Air data transducer assembly and computer; safety valve (cabin air pressure); solenoid valve (shutoff, air)
Albany International Company (FRL) Dedham, Massachusetts	Nomex felt (for the thermal protection system)
Ball Aerospace Systems Division Boulder, Colorado	Star tracker
Bendix Corporation (Navigation and Control Group) Teterboro, New Jersey	Airspeed altimeter; vertical velocity indicator; surface position indicator
The B.F. Goodrich Company Akron, Ohio	Main and nose landing-gear wheel and main landing-gear brake assembly
Boeing Aerospace Company Seattle, Washington	Carrier aircraft modification
CCI Corporation (The Marquardt Company) Van Nuys, California	Reaction control system thrusters

(*Table 8-2. continued overleaf*)

Table 8-2. continued

Company	Product
Conrac Corporation West Cardwell, New Jersey	Engine interface unit (main propulsion system)
Corning Glass Works (Technical Products Division) Corning, New York	Windows, windshield and side hatch window; glass-ceramic retainers (for thermal protection systems tiles)
Cutler-Hammer (AIL Division) Farmingdale, New York	Microwave scanning-beam landing-system navigation set
Fairchild Republic Company Farmingdale, New York	Vertical tail
General Dynamics Corporation (Convair Division) San Diego, California	Mid fuselage
Grumman Aerospace Bethpage, New York	Shuttle wings
Honeywell (Avionics Division) St. Petersburg, Florida	Flight-control system displays and controls
Hughes Aircraft (Space and Communications Group) Los Angeles, California	Ku-bank radar communication system
Hydraulic Research Textron Valencia, California	Servo actuator elevon-electro command hydraulics; fourway hydraulic system flow control pressure valve

DOMESTIC SPACE BUSINESS 109

IBM (Federal Systems Division) Oswego, New York	Mass-memory and multifunction cathode-ray-tube display subsystem; general-purpose computer and input-output processor
Instrument Systems Corporation (Telephonics Division) Huntington, New York	Audio distribution system (voice and tonal signals)
Lear Siegler (Instrument Division) Grand Rapids, Michigan	Attitude direction indicator
Lockheed-California Company Burbank, California	Static and fatigue testing of orbiter structure
Lockheed Missiles and Space Company Sunnyvale, California	High- and low-temperature reusable surface insulation
Martin Marietta Corporation New Orleans, Louisiana	External tank
McDonnell Douglas Astronautics Company Huntington Beach, California	Solid rocket booster structure
McDonnell Douglas Corporation St. Louis, Missouri	Aft propulsion system
Menasco Burbank, California	Main and nose landing-gear shock struts and brace assembly
Northrop Corporation (Precision Products Division) Norwood, Massachusetts	Rate gyro assembly

(Table 8–2. continued overleaf)

Table 8-2. continued

Company	Product
Rockwell International (Rocketdyne Division) Canoga Park, California	Space shuttle main engine
Rockwell International (Space Systems Group) Downey, California	Space shuttle orbiter system integration
Rockwell International (Tulsa Division) Tulsa, Oklahoma	Cargo-bay doors
Spar Aerospace Toronto, Ontario	Remote manipulator system
Sperry-Rand Corporation (Flight Systems Division) Phoenix, Arizona	Automatic landing; multiplexer-demultiplexer
Sundstrand Corporation Rockford, Illinois	Auxiliary power unit; rudder-brake actuation unit; Actuation unit (body flap) hydrogen recirculation pump assembly (main propulsion system)
Thiokol Corporation (Wasatch Division) Brigham City, Utah	Solid rocket booster motors
TRW (Electronic Systems Division) Redondo Beach, California	S-band payload interrogator; S-bank network equipment; Network signal processor; payload signal processor

United Technologies Corporation (Chemical Systems Division) Sunnyvale, California	Solid rocket booster separation motors; propulsion for inertial upper stage
United Technologies Corporation (Hamilton Standard Division) Windsor Locks, Connecticut	Atmospheric revitalization subsystem; freon coolant loop and flash evaporator system; water boiler, hydraulic thermal control unit; shuttle spacesuit
United Technologies Corporation (Power Systems Division) South Windsor, Connecticut	Fuel-cell powerplant
United Technologies Corporation (United Space Boosters, Inc.) Sunnyvale, California	Solid rocket booster assembly (checkout, launch)
Vought Corporation Dallas, Texas	Leading-edge structural subsystem and nosecap, reinforced carbon-carbon; cargo-bay door radiator and flow-control assembly system
Westinghouse Electric Corporation (Aerospace Electrical Division) Lima, Ohio	Remote power controller; electrical system inverters
Westinghouse Electric Corporation (Systems Development Division) Baltimore, Maryland	Master timing unit

Source: From *The Space Shuttle Operator's Manual*, by Kerry Mark Joels, Gregory P. Kennedy, and David Larkin. Copyright © 1982 by Kerry Mark Joels, David Larkin and Gregory P. Kennedy. Reprinted by permission of Ballantine Books, a division of Random House, Inc.

Table 8-3. Selected Major Aerospace Corporations and Area of Expertise.

Corporations	Satellites	Rockets[b]	Data Processing and Computers	Other Research and Instruments	Astronaut-Rated
Aerojet General[a]					
Ball Aerospace[a]	X				
Bell Aerospace Textron[a]		X			
Bell Labs					
Bendix Corporation[a]	X			X	
Boeing Aerospace Company[a]		X		X	X
Chrysler		X		X	X
Computer Sciences Corporation			X		
Draper Laboratory				X	X
Fairchild Space and Electronics	X	X		X	
General Dynamics[a]		X			
General Electric[a]	X			X	
Goodyear Aerospace[a]			X	X	
Grumman Aerospace[a]				X	X
Hughes Aircraft[a]	X				X
IBM[a]			X		
Litton Industries		X			
Lockheed Missile and Space[a]		X		X	X
Martin Marietta[a]	X	X		X	X
McDonnell Douglas[a]		X		X	X
United Technologies[a]		X			X

RCA[a]			X	X	X
Rockwell International[a]	X	X			
Thiokol Corporation[a]					X
TRW[a]	X		X	X	
Vought Corporation[a]	X	X			
Westinghouse[a]	X	X			

a. AIA member.
b. Boosters such as the Saturn 5 and the external tanks of the Shuttle, although astronaut-rates, are included under rockets.

Source: Table in Nathan Goldman and Michael Fulda, *U.S. Space Policy*, manuscript under preparation. Data from CRS, "United States Civilian Space Programs 1958–1978," Subcommittee on Space Science and Applications, the Committee on Science and Technology, U.S. House of Representatives, 97th Cong., 1st sess., January 1981, pp. 935–73.

Table 8-4. 1979 Statistics for Selected Aerospace Companies.

Company	Fortune 500 Rank	DOD Contracts ($ millions)	NASA Contracts ($ millions)	Total Employees
Boeing Aerospace Company	29	1,515	101.5	98,300
General Dynamics	83	3,492	46.9	81,600
Grumman	216	1,364	6.0	28,000
McDonnell Douglas	54	3,229	113.7	82,700
Northrop	204	800	26.1	28,800
Rockwell	45	684	1071.8	114,452
United Technologies	26	2,554	73.3	197,700

Source: From Gordon Adams, *The Iron Triangle: The Politics of Defense Contracting* (New York: Council of Economic Priorities, 1981); *Aerospace Facts and Figures*, AIA, 1982–83.

The company has to decide whether to build the components itself or to subcontract. The major companies have often bought out specialty houses so that they could offer that talent in the balance when bidding for the prime contract. Table 8-3 includes twenty-eight aerospace companies and shows the diverse and overlapping expertise of each. Note that twenty-one of the twenty-eight belong to the Aerospace Industries Association, a trade association.

In addition to vertical integration within the industry, these companies have diversified beyond the aerospace industry. TRW, for instance, is involved in automobiles, electronics, and energy as well as aerospace. In 1981 it was valued at over $5 billion and was seventy-first in sales. Table 8-4 lists some major aerospace companies. Notice their rank among the largest corporations in the nation. The table also depicts the cyclical nature of the aerospace contracts: Rockwell still rode at the crescent of its boom with the Shuttle contract, but Grumman wallowed in the trough of the cycle. The dominance of defense over civilian space contracts, likewise, is apparent from this table. Of these seven corporations, only Rockwell received more NASA contract money than DOD money.

As shown in Table 8-4, more subcontracting companies are located in California than in any other state. The Pacific Coast is the home of the aerospace industry. Located in the Pacific states is 26.2 percent of the military aircraft and 41.3 percent of the civilian aircraft employment. The story is the same for missiles and space. Located in those states is 41.2 percent of missile and a whopping 62.3 percent of the space employment.[5] Only the NASA policy of distributing contracts for geographic dispersal keeps aerospace manufacture from being even more lopsidedly West Coast.

TRADE ASSOCIATIONS AND LOBBYING

The strength of the aerospace industry can be overstated: The General Accounting Office is a watchdog agency of the government; DOD and NASA have both caught some companies overcharging the government and have fined them heavily. Gansler argued in this respect that although the defense industry experienced massive cost overruns, two-thirds of these overruns were the result of government changes in contract and rising inflation. Moreover, DOD had done better than the Departments of Transportation and of Energy or the Veterans Administration in limiting cost overruns.[6]

Nonetheless the aerospace industry does have a great deal of political clout. Despite statutorily set periods between a person leaving government and entering industry, a strong old boy/old girl network exists among the Congress, the DOD/NASA, and the clientele companies.[7] These companies all have Washington offices with a lobbying or information function. The industry also has a very active set of political action committees (PACs).

Finally, the industry has a very active set of trade associations. The Aerospace Industries Association (AIA) is located in Washington, D.C. and is a registered lobby (although most active lobbying is done by the individual companies). Forty-eight corporations belong to the Association. AIA is a member of the Council of Defense and Space Industry Association (DOCSIA). These groups constitute a powerful force in Washington.

In addition, several new trade associations have sprung up in the last few years that deal specifically with space concerns. The International Association of Satellite Users represents satellite users before the FCC at home as well as before the ITU abroad.[8] The Society for Private and Commercial Earth Stations (SPACE) and the National Satellite Cable Association represent the interests of the private cable operators and the C-band station dealers and distributors. The Direct Broadcast Satellite Association was formed in 1983.

COMMUNICATIONS INDUSTRY

The other great beneficiary of space commerce has been the communications industry. Satellites have provided more capacity at cheaper prices than traditional cable and wire (optical fibers may reverse this balance). This capacity and price have facilitated an explosion in applications of communications, including electronic mail, teleconferencing, teletext, and satellite television.

The annual Satellite Industry Directory issue by *Satellite Communications*, a leading technical magazine, depicts the growth of the satellite industry. Between 1981 and 1982 the number of satellite carriers listed rose from sixteen to twenty-five. The communications big names were well represented: Comsat, GTE, AT&T, Hughes, RCA, and Western Union. The satellite users, especially the cable users, are not included in the directory, but these companies and services have become household words also because of satellites: HBO (Time, Inc.), MTV, and CNN Turner Broadcasting System (TBS) have each revolutionized a portion of the communications industry.

The number of service companies, including trade associations and publications, consultants and resale carriers, has increased from about 120 in 1981 to over 200 in 1982. The hardware companies listed in the directory, however, have remained constant at over 300 firms. By 1983, the directory lists over 600 companies involved in the satellite communications industry.

The communications market, however, shows signs of change in the middle 1980s. The FCC actions of the early 1980s lead to a temporary excess of transponders capacity. Also, cable television now offers fifty video choices, but only HBO and Showtime are making money. Even MTV is still losing money. The recent merger of Cable Health Network and Daytime probably suggests the wave of the future. But as the cable television industry undergoes its eventual winnowing out of weak companies, the industry has entered a new era of expansion in the business communications market with teletext, electronic mail, and teleconferencing. Direct broadcast satellites also will continue to provide new business to the satellite communications industry.[9]

OTHER PARTICIPANTS

Consultants

Another established industry is technical and financial consultants. In addition to their in-house experts, business and government are accustomed to hiring outside analysts to assess programs and even to forecast future scenarios for planning purposes. Booz, Allen, and Hamilton; Frost & Sullivan; and the National Academy of Public Administration are among the groups that have conducted substantial studies on space under contract. Arthur D. Little, a major management consulting firm, is also one of the premiere consultants on space commerce. Similarly, Rand Corporation has expertise on the military space applications. Increasingly, the major accounting firms have begun to enter the consulting field, including the new space area. More recently, the Center for Space Policy has been incorporated as a consulting firm for industry and government.

Another set of consultants handle the engineering aspects of space projects. Other firms handle space law and space policy as well as investment and regulatory consultants. Some of these consultants are university professors. Other consultants have gained expertise in aerospace companies and now freelance as space consultants. This group of consultants is most common in California. Space consulting firms are often headquartered in Washington, D.C. but increasingly they exist in space centers such as Florida, Texas, and California.

Space Insurance

Another business created by the boom in space commerce is space insurance. In the early days of space satellites, the ever-risky space flights were insured by Lloyds of London. By the 1980s major insurance companies were providing insurance for satellites. The seven insurance brokers of space business in the United States are Alexander and Alexander; Corroon and Black Inspace; Frank B. Hall; Johnson and Higgins; Marsh and McLennan; Nausch, Hogan and Murray; and Reed Stenhouse.

After the space client has selected a broker, the broker approaches the underwriters for a rate quotation. The client then decides which quotation, usually the lowest, to accept. The broker then presents the decision to the underwriters, who each decide whether to participate at this rate and risk of course. The rates change rapidly based on the new technologies and experience with satellites. Prices are much more unstable because no established actuarial experience has been accumulated concerning space. MBAs, lawyers, political scientists, and engineers are employed by these companies to conduct their multidisciplinary analyses. At least seven forms of space insurance can be purchased: prelaunch, launch, in-orbit satellite life, transponder life, service interruption, special event, and third-party liability insurances. The leading underwriters in space risk are American Foreign Insurance Association; Intech; Lexington Insurance Company; Montgomery Collins; R. W. Bryant and Company; Stewart Smith East; and USAIG Satellite Operations.

The market in insuring satellites for the 1980s has been estimated at $220 million in commission for the estimated 200 satellites.[10] The loss of several satellites from 1979 to 1981 foreshadows the risk of the industry. Rates rose considerably, and the failure of two satellites in February 1984 rendered the market forecast obsolete.[11] It should also be pointed out that U.S. insurance companies have also insured the Ariane for loss. It may seem ironic that an American firm is insuring a foreign competitor, but the fact of international business is that if a U.S. firm does not do it, someone else will. The commissions are paid to American brokers, not to the foreign companies.[12]

Banking and Financial Institutions

The major banks can make or break a company or program. The audits and resulting opinions determine whether the banks will issue loans to the company; these decisions also affect the stock and debt of a company. The major banks that have been involved in loans to the aerospace industry are Chase Manhattan, Citicorp Bank of New York, Bankers Trust, Manufacturers Bank, and Mellon Bank. Manufacturers Hanover Trust, for instance, helped finance U.S. Satellite Systems, a new space communications venture.[13]

Accounting Firms

In the last few years the "big eight" accounting firms have begun to broaden their activities to include direct consulting. Several firms have begun to develop such an expertise in space commerce. Coopers and Lybrand has established itself as the leader among accounting firms in space commerce. It has consulted companies in DBS and other communications endeavors, companies in launch vehicles, and in space manufacturing. Coopers and Lybrand has one of the two contracts from the NASA space station task force; Booz, Allen, and Hamilton, the consulting firm, has the other contract.

Among the other big eight firms, Touche Ross has done work with cable television. Recently both Arthur Young and Peat, Marwick, and Mitchell have begun to enter the space consulting field.

Space Lawyers

Another segment of the growth industry has been space law. For more than a quarter century, space law has been an esoteric realm of international diplomacy and treaty making (see Chapter 9 on international space law). In the 1980s, space law, in both its domestic and its international aspects, has become more important. Courses on space law have begun to appear in law school curricula. George Washington University and the University of Mississippi each offer a course yearly. Many California law schools, the location of nearly half of the space effort and industry, also offer courses on space law.

Moreover, the American Bar Association (ABA) has begun to recognize the importance of space law. In 1981 the American Law Institute (ALI) and the ABA conducted the International Conference on Doing Business in Space. The Second Institute on Aviation and Space Litigation, in 1983, held a panel on commercial space activities. The ABA has a Subcommittee on Space Law in its section on international law. It also has a Committee on Outer Space Insurance in the insurance section and a Committee on Outer Space in the science and technology section. The Federal Communications Bar Association in Washington, D.C. has become extremely active in space law because of the expansion in satellite communications since the late 1970s.

International space law, relatively esoteric, has long been the domain of legal scholars and legal diplomats. Lawyers in NASA, the U.S. Departments of Defense and State have negotiated, debated and written about the space law treaties and agreements. The aerospace companies also maintain in-house counsels who concern themselves with space law.[14]

Space industrialization has recently expanded both the number of lawyers and the type of law encompassed by space law. Contracts, patents, and property have become part of this new regime. Among the important Washington law firms heavily involved in space is Schnader, Harrison, Segal, and Lewis. Although an all-service law firm of 165, it has a special high-technology and space group that includes Paul G. Dembling, former General Counsel of NASA, and Delbert D. Smith, author of *Space Stations: International Law and Policy* and other space books and articles. In the law firm of Dickstein, Shapiro, and Morin, Leigh Ratiner served as lobbyist and lawyer for the L-5 (Space) Society and successfully led the effort in 1980 to have the Senate Foreign Relations Committee table and kill the proposed Moon Treaty.

Another example is the Washington-Dallas-Houston law firm of Andrews and Kurth. James Myers is the resident expert on space law. This major firm is the counsel for SSIA and SII. Eagle Engineering, which contracts with government as well as with SSIA and SII, is represented by Arthur Dula, a high-tech lawyer in Houston and cofounder of Eagle Engineering.

Attorney Dula exemplifies the new space lawyers. He has written law review articles on both domestic and international space law. As a member of the International Institute on Space Law, he presents papers at yearly colloquia. He is active in space interest groups and space businesses, organizing the 1983 Second Annual L-5 Conference on Space Development in Houston. In addition to his part ownership of Eagle Engineering, Dula incorporated SSIA and approximately twenty-five other high-tech companies. He served as SSIA's general counsel until it became too large and was reincorporated in Delaware. He now serves as its aerospace law counsel.

NEW USERS: NEW PRODUCERS

The space commerce market has entered a new era of entrepreneurial ferment. The conservative nature of existing firms must be pointed out: The risk-taker is rare but once a risk-taker has proven a new market or product, others will follow quickly. Johnson and Johnson, with McDonnell Douglas, pioneered the space pharmaceutical field. The results of the Shuttle experiments have been very promising. Evidence already is mounting (although details are proprietary with the companies) that other pharmaceutical companies, such as Eli Lilly, are planning to enter the space pharmaceutical market in the next several years. The other manufacturing areas—semiconductors and alloys—are not as near to production, so the industry has yet to respond with as many entrants into the field. According to a senior vice president of Booz, Allen, and Hamilton, almost seventy nonaerospace companies were investigating the opportunities offered by space manufacturing. Seventeen venture capital firms also were investigating space commerce for potential investment.[14]

Anatomy of a New Space Company

These new concerns involve new ideas or new applications of old ideas—ELVs, the Shuttle, Comsat, remote sensing, and private space stations. Each requires a creative entrepreneur: a Harvard Business School MBA for Orbital Sciences Corporation; a former NASA chief engineer for Space Industries. These new companies start out as small concerns with a handful of engineers designing the hardware.

Yet at an early stage, the concerns become financing the project and wading through the government's red tape. To finance a project requiring tens of millions to billions of dollars, the space entrepreneur must become an excellent salesperson. The first investors are needed to provide the money to complete the planning and prototypes. SSIA, for instance, raised more than $6 million from fifty-seven investors before its Conestoga launch of September 1982. These investors were mostly Houston oil and real estate people. Microgravity Research Associates has raised $3 million mostly from Texas oil and banking interests.[15]

Orbital Sciences, the OTV Company, first raised money through an investors' consortium including Rothchild, Brentwood Associates of Los Angeles, and others. It became a limited partnership in late 1983, and issued a private offering for a thousand $50,000 units. The offering, through Shearson/American Express, sold almost one-half of the units within two weeks.[16] Space Industries also planned a private offering in 1984.

But the big investment money must come from elsewhere. One source has been the insurance industry. For instance, Aetna Life Insurance Company owns one-third of Satellite Business Systems (SBS). Another company heavily involved in space business has been Prudential Insurance. Although the deal did not succeed, Prudential was reported to have been negotiating to buy a 40 percent interest in the Space Transportation Company (the fifth orbiter). Prudential did eventually acquire a large interest in United Satellite Communications, the first DBS company, along with a coinvestor, General Instrument Corporation.[17]

Another example that demonstrates several points already made is the case of Transpace Carriers, which wants to purchase the Delta booster from McDonnell Douglas. The company was founded by

David Grimes, who managed NASA's Delta Team for many years. By late 1983 the company had a dozen employees and had gained the financial backing of Cigna, which is the holding company for the Connecticut General Life Company and the Insurance Company of North America. Other financial backing—in another example—can come from investment groups such as the Swedish Stenbeck Group, a part owner of Sparx, the multinational remote sensing concern.[18]

Another source of support may be a consortium with an established company in a related area. The Space America consortium, along with Bendix Corporation, made a bid in 1983 to take over Landsat as a private endeavor. The other bidders were Comsat General Corporation (then a partner in Sparx) and RCA Astro-Electronics (the private contractor for Landsat).[19]

Many new companies, however, never get this far, because they are chekmated by the actions of established firms. SSIA bowed out of the competition for the ELV Atlas-Centaur, when the Centaur's manufacturer (General Dynamics, Convair) decided to create its own Commspace Department and market the craft itself.[20] Similarly, the Orbital Sciences Corporation's plans for a commercial upper stage are receiving stiff competition, in part because United Technologies is selling upper stage rockets directly to the satellite companies, effectively bypassing OSC.[21]

Using the computer industry of the 1970s as an example, many new space companies will be bought out by established aerospace or communications companies who already have the capital or the technical expertise but who want the creative ideas of the new space company. A minor version of this process appeared in 1983 when Federal Express bought the name of the Space Transportation Company as well as its rights to negotiate with Martin Marietta for the marketing rights for the Titan 34-D ELV. Although the negotiations proved unsuccessful, Fedex Space Tran continues to exist and may develop the new space link in Federal Express to deliver electronic mail by satellite.[22]

The Inevitable Shakeout

Some people are going to be overoptimistic in their analysis of the business conditions. Others may be right, but conditions may change rapidly in the technologically expanding fields. Which companies will provide the space IBMs of the 1990s may be more a job for a fortune teller than for a securities analyst.

The big aerospace corporations may be the most likely winners because of their financial and technical resources. They can afford miscalculations more than the small companies, and they are more likely to form alliances with companies at home and abroad. Moreover, these companies are likely to dominate corporate mergers with the new entrepreneurial space companies.

The commercialization of outer space has created an entire new set of industries and revitalized an established one. The established aerospace companies, which have been victims of a cyclical industry, may find their new business as a way out of this cycle of insecurity. New services have sprung up in communications, revitalizing the television and telephone industry, wedding satellites and computers, and creating whole new forms of communications. Remote sensing and space manufacturing are on the verge of fulfilling decades of promise as new industries. The new decade should bring new companies, new products, and new creativity in space, invigorating an important sector of the economy.

But new endeavors can bring unanticipated problems; the government must be ready to deal with these problems. Unfortunately, the governments of the world have heretofore been slow to respond to the new frontier and its social, economic, and political challenges.

NOTES TO CHAPTER 8

1. The primary exception was communications satellites.
2. *Aerospace Facts and Figures*, Aerospace Industries Association, 1982–83.
3. Jacques S. Gansler, *The Defense Industry* (Cambridge, Mass.: MIT Press, 1980); Charles D. Bright, *The Jet Makers* (Lawrence: Regents Press of Kansas, 1978); Barry Bluestone, Peter Jordon, and Mark Sullivan, *Aircraft Industry Dynamics* (Boston: Auburn House, 1981).
4. Gansler, *The Defense Industry*, pp. 128–137.
5. *Aerospace Facts and Figures*, pp. 132–33.
6. *Ibid.*, pp. 90–93.
7. Gordon Adams, *The Golden Triangle: The Politics of Defense Contracting* (New York: Counsel of Economic Priorities, 1981), pp. 57–61.
8. *Space Age Review*, September 1981.
9. *Satellite Communications*, September 1983.

10. Ann Deering and George R. Prado, "Doing Business in Space," *Space Age Review*, September 1981.
11. Conversation with Oliver Hennigan, Jr., space analysis for USAIG, 1983 — exact rates are confidential.
12. Andrew Tobias, *The Invisible Bankers: Everything the Industry Never Wanted You to Know* (New York: Linden Press, 1982).
13. Adams, *The Iron Triangle,* pp. 65-71, *DBS News,* December 1983.
14. Author interviews—aerospace lawyers Arthur Drula, James Myers, and Edward Finch, Jr.
15. *High Frontier Newsletter,* December 1983.
16. *Austin American Statesman,* January 17, 1982, September 12, 1982, October 12, 1982; *Aviation Week and Space Technology,* September 13, 1982, February 15, 1982, June 14, 1982, September 6, 1982; *Houston Post,* September 21, 1982; *Los Angeles Herald Examiner,* September 25, 1982.
17. *Commercial Space Report,* December 1983, *Space Business News,* July 18, 1983.
18. *DBS News,* November 1983, *Space Business News,* July 18, 1983.
19. *Satellite Communications,* November 1983.
20. *Space Calendar,* November 21-27, 1983.
21. *Ibid.,* June 20-26, 1983.
22. *The Commercial Space Report,* November 1983.
23. *Space Calendar,* September 5-11, 1983, October 24-30, 1983.

9 U.S. POLICY TOWARD SPACE COMMERCE

The U.S. government finds itself with a very delicate role to play in world space commerce. The United States must alternately cooperate and compete with foreign space powers both nationally and in collaboration with business corporations. Inherently that demeanor puts the U.S. government in conflict with its domestic private enterprises. This complex situation will require some agile policymaking by NASA and the other space planners in government.

COOPERATION OR COMPETITION

The policy of the U.S. government in all fields is to avoid competition with private domestic industry. As long as space concerns had been largely national security and experimental, the problem really did not arise. When it did arise in the communications field, the government reached an agreement quickly with a new public-private solution—Comsat, and then with the government-regulated monopoly—AT&T. Telecommunications was already part of a long-regulated industry. In the 1960s, the development of this private space communications was part of a national policy, generally orchestrated by the government. Yet the push for private development of space in the 1980s comes more directly from private entrepreneurs. The government now is the one having to form a policy post hoc.

Communications

In communications, the consensus in the early 1980s supports the deregulation of the industry to permit competitive activity. The government hopes that such activity will be creative and help produce the R&D and managerial improvements to permit the continued leadership of the United States in science and engineering so important for economic, military, and political power today.

In communications, Congress is not writing on a *tabula rasa*, however. The hoary Communications Act of 1934, written in the depths of the Great Depression, attempts to regulate communications as if it were still a monopolistic sector. Operating under this restraint, the last four administrations, through the FCC, have been deregulating the communications industry, including satellites. The Federal Trade Commission (FTC) and the Justice Department also pursued this policy in the 1983–84 break-up of the AT&T conglomerate.

Congress has attempted, less successfully, to deregulate the telecommunications industry. In 1981 Senator Robert Packwood (R–Ore.) introduced the Telecommunications and Deregulation Act, S.B. 898. AT&T supported the bill, but GTE (often an ally of AT&T) and the International Communications Association did not approve, on the grounds that the market was insufficient to permit free competition. Other established industries feared the spillover of an unregulated market, and particularly an unregulated AT&T. Newspapers feared satellite-transmitted "electronic yellow pages" would reduce their advertisement revenues in an already competitive market. The National Association of Broadcasters (NAB) likewise feared the subsidized competition from syndicates in their local markets.[1]

In the House of Representatives, Timothy Wirth (D–Colo.) cosponsored the proposed Wirth–Broyhill Communications Act of 1982, H.R. 5182. This bill would have partially deregulated the industry. American Telephone and Telegraph would be designated as a "dominant carrier" and continue to be regulated as well as audited annually by a certified public accountant. AT&T initiated a $2 million lobbying campaign to halt the bill. In light of the court-approved breakup of AT&T, the bill could be portrayed by the industry as an unnecessary complication.[2] Similar legislation in 1983 was also not passed by the Congress.

Remote Sensing

The reticence of the policymakers to make definitive decisions on space-related issues is repeated in the remote sensing sector. In 1981 the Department of Commerce (DOC) drafted a bill to regulate the new industry. Title I would require DOC to license the private company. Earth station locations and broadcasting frequencies would need to be allocated.

Under Title II, the Secretary of Commerce could contract with these new companies or the company managing Landsat in order to purchase data for the government. Other articles of the bill would have NOAA to operate the Landsat until private ownership. Moreover the government could enter into ventures with private industries for both R&D and for operations.[3]

This bill, along with those to transfer Landsat and weather satellites, has met skepticism and opposition and has little chance of passage. Yet it is a testament to the potential of this market that so much technical and entrepreneurial development has taken place in this turmoil.

President Reagan made a political mistake when he combined the privatization of weather satellites with the Landsat satellites. Weather forecasting is generally seen as a national resource, and the suggestion activated a lot of fears. Likewise, from an economic standpoint, very little in the way of a private market has been isolated. In late 1983 the House voted 377 to 28 to disapprove the sale of weather satellites.[4] Finally, a proposal to sell Landsat, monopolistically attractive until 1981-82, has been confronted with several competitors that may be more effective in the marketplace, by adjusting their own hardware and marketing strategies to more lucrative segments of the market. Landsat, after all, is a government R&D, proof-of-concept project and simply may not be competitive with private or foreign alternatives.

In 1984, the House approved a new format for Landsat. NOAA would retain ownership, but a private company would be selected to operate and market the data. For later generations of remote sensing satellites, the government would contribute some subsidy but the new satellites would be predominantly private enterprises. Concurrent with this legislative action, the Department of Commerce (DOC,

NOAA) proceeded with its request for proposal to sell those marketing rights to private industry. In April 1984, the DOC selected Space America, Kodak, and RCA as the three finalists for the rights.

Transportation

The U.S. space bureaucracy of the middle 1980s is a thirty-year hodgepodge of overlapping missions and agencies. NASA has a broad but vague mandate from the 1958 National Aeronautics and Space Act to maintain the American lead in space technology. NASA was established as an R&D agency; the need for a creative R&D program in a competitive world remains a strong policy reason for maintaining NASA in that capacity. In this regard, the establishment of a Comsat and the transfer of Landsat to NOAA's control reflect the thinking that NASA should not be encumbered by an operational orientation.

NASA nevertheless has maintained both an operational dimension and a regulatory dimension. The recent decision by NASA to discontinue ELVs and rely completely on the shuttle reflects that outlook as well as the assessment that redesigning the satellite for an expendable was more expensive and more time-consuming than waiting for the next available shuttle. Permitting private industry to acquire the ELVs, however, raises the question of who will regulate the launching of these rockets. NASA has done some regulation before: Until Congress extended U.S. criminal jurisdiction to space, NASA had published rules in the Federal Register to authorize the Shuttle commander to have the powers of a ship's captain over his or her crew and passengers. Moreover, NASA regulates the contents and procedures for satellites and other users of the Shuttle or expendable vehicles.

Nonetheless, the growing uses of space and the rise of a separate space industry render this NASA ad hoc regulation outmoded. The SSIA example demonstrates the multiple agency involvement inherent in the present arrangement. In response, proposed legislation have been offered in both the Houses of Congress to provide for a lead agency to regulate the new private concerns in space. Some bills proposed to have the FAA act as the lead agency. First proposed by Senator Howard Cannon (D–Nev.) in 1982, the FAA would seem the best candidate for regulating private space launches; the duty is not that unlike its more earthly (airbound) duties.

Other bills, however, propose vesting the lead agency mantle with the Department of Commerce. In the longer view of space commerce, the DOC seems to be a much better seat of authority. Commerce or even NASA are better situated to provide regulations for space manufacturing and the other exotic new space industries.[5]

This bill was cosponsored in 1983 by more than fifty members of the House, including conservative Jack Kemp (R-New York) and liberal Mo Udall (D-Arizona). The DOC would license and regulate private launching. After formal notice and a public hearing, the Secretary of Commerce would issue or deny the license as well as attach "conditions" such as insurance requirements or national security considerations on the flight. At the end of 1983, President Reagan followed the advice of the Senior Interagency Group's study on space policy and recommended that the Department of Transportation (DOT) be the lead agency for both licensing the private vehicles. Moreover, when the Space Shuttle becomes fully operational, the Department of Transportation may manage the Shuttle fleet as well. Although the original proposed legislation in 1983 (Appendix H) stipulated that the Department of Commerce be vested with ELV licensing authority, the latest version of this pending bill would place the licensing function in the hands of the DOT.

NASA would remain free to provide the basic research and the new programs that will allow the United States to maintain its leadership in space. NASA would be able to construct space stations and continue to nurture the development of space commerce, especially space manufacturing. Moreover, as the president's Science Advisor, George Keyworth, has advocated, NASA and the United States should begin to look beyond the short term on earth and in space. Projects that will further the U.S. position in space commerce and in other aspects of space include the heavy life vehicle to transport raw materials and other supplies into space; a permanent space station as a site for space research, manufacturing and construction; an astronaut-rated orbital transfer vehicle to carry workers to geosynchronous orbit; space construction techniques for large space structures (LSS) such as antenna-farm satellites, solar powered satellites and space lasers; finally, a lunar colony that could mine the moon for ores and oxygen (fuel) as well as be a science outpost. While Congress or the president must provide the vision, a competent research and development agency such as NASA must be in a position to fulfill that mission.

Manufacturing

None of the proposed bills on space commerce deals directly with the regulation of space manufacturing. The proposed "Space Commerce Act" of 1983 did include Condition (4), requiring the Secretary of Commerce to "approve the proposed payload" when it licensed a private launch. This provision would cover private free flyers that contain materials-processing facilities. Even so, the provision is just not adequate.

Nor has Congress dealt with the role of a NASA space station in the commercialization of space. NASA needs to be given (or to develop) guidelines to permit private access to its facilities in space. At present NASA is concerned with creating new potentials in space and has been generous with its JEAs and TEAs. As space processes become better developed, however, the policymakers will need to decide how to use the publicly developed space resources to the benefit of society. Someone will have to decide the priorities of access to space businesses in terms of the prices and availability of services and the degree of exclusivity (monopoly) granted in certain processes.

The Lead Agency

Lurking behind all these policy decisions is the question of who will make and implement these space policy decisions, although the ultimate decision need not be answered all at once. In the early days of the U.S. aviation industry between 1920 and 1940, the Congress had to modify the administrative framework on average every two years.[6]

A similar process is already apparent in the proposed and passed legislation dealing with space in the 1970s and 1980s. Remote sensing presently is situated under the jurisdiction of the Department of Commerce. Communications is regulated by the FCC and some R&D is conducted by NASA. As of 1983, NASA also operated the transportation and manufacturing sector. After a fruitless jurisdictional struggle between the FAA and DOC, it is now the Department of Transportation that has been designated as the agency in charge of licensing and regulating private ELVs and possibly the Shuttle itself. No legislation to date, however, has dealt with the space manufacturing sector. NASA will do this task as well as operate the Shuttle by default, but only at the sacrifice of its prime mission as the lead research and development agency for U.S. civilian space efforts.

The present environment seems to favor greater segmentation of the space sectors. Although this process has its advantages, especially for a revitalized NASA, fragmentation may be detrimental to long-term, consistent space policy. The Reagan administration has been working on an overall space policy for three years, but the rapidly changing situation continues to confound its efforts.

A PROGNOSTICATION

The trial-and-error path of administration will eventually lead to a Department of Space, probably not before the 1990s. Elsewhere, I have explained need for and purpose of such an agency and suggested a general structure for it.[7] Today more than a dozen agencies have some space-related duties. As space becomes more intrusive in the U.S. economy, the tendency will be for more agencies to become involved in space activities as users or regulators.

This proliferation of actions and actors will make unified space policy much more difficult in the face of growing world competition. A Department of Space would be able to coordinate space activities. It would absorb most of NASA and would be divided into offices headed by assistant secretaries of research, operations, and planning.

The exact duties of a Department of Space will develop by trial and error. Only experience can identify which space activities share the same fundamentals and can be managed in the same organization and which activities should be integrated instead according to activity. For instance, the FCC will continue to regulate all communications, space- and earth-based. On the other hand, whether the Occupational Safety and Health Administration (OSHA) or the Department of Space will regulate safety standards in private and public space stations (factories) is a question for the future.

Another problem of a hypothetical Department of Space is a more focused version of the problem of coordinating a private and a public space effort. If the same body regulates private space manufacturing or space launching and operates its own launch and manufacturing facilities (ELVs and the Shuttle today), can that body be trusted to regulate its own competition?

Since the Department of Space is a speculative proposition at best, let us return to a more near-term, feasible alternative for coordinating U.S. space policy. This could be in the executive branch, a reestablishing National Space Advisory Council (NASC), and in Congress, reestablishment of Space Committees. In the early 1970s, the NASC was abolished and the Congress's Space Committees were downgraded to subcommittees within other committees. This decision reflected the mood of the period: Apollo was over; foreign competition (indeed the military and economic importance of space) was not fully conceived. Yet in less than ten years, the situation had changed radically. The economics of space was becoming a multibillion dollar industrial sector. And although beyond the scope of this book, space had obtained the military importance predicted for it since the 1950s.[8]

By 1984 the United States continued to struggle without a coordinated, high-level space policy. Both Presidents Carter and Reagan ordered space policy studies. The plan suggested by Carter's study to coordinate military and civilian space efforts created intense opposition in the late 1970s. Many feared that the military would take over the weakened NASA, and the study led to no action.

By the 1980s, however, private space activities had strengthened the civilian hand in space. Now may be the time to create an expanded NASC. This Council would contain the secretaries (administrators) or assistant secretaries of NASA, DOD, and NOAA—the operational space agencies. NASC would also include representatives from the users of space resources—the Departments of Agriculture, Interior, Commerce, and others. To quiet further the fear of military or even government domination, representatives of the space industry (the AIA) and labor (the IAMAW) could also be included in the council.[9]

The goal remains to provide a deliberative body that can coordinate space policy for the United States so that this country will be prepared to meet the world competition in the commerce, politics, and military aspects of space exploration and exploitation.

COMMERCIAL SPACE POLICY

Role of the Government

In addition to providing the structure for space policy, the government is responsible for the substance of space policy. The U.S. government has a primary goal to maintain U.S. leadership in space, both politically and economically. The corollary of this goal is to acquire private enterprise as a partner in pursuing this goal.

This strategy provides the government with a set of functions. The government initially created the infrastructure of space development and must continue to invest in the research to maintain that infrastructure. This role is especially important since the competing space powers, the Soviet Union, Europe and Japan, have programs which are controlled by or heavily subsidized through government research and funding. Coordinating potentially conflictual military, civilian and private space programs will become increasingly difficult but increasingly necessary. The government's most important role, however, may be the continued development of advanced research and development. In remote sensing (the Thematic Mapper), in communications (30/20 GHz), in transportation and manufacturing, the government must take the lead in developing the technology for maintaining the U.S. position in the world.

The U.S. government still should identify which space activities are better conducted by private industry. It also needs to help the private space endeavors through the morass of government regulations. Similarly, the government must assess its activities in its own sector to make sure its policies (such as the pricing of the Shuttle) do not cripple the private effort. More often, however, the nonactions of the government have harmed the private space effort.

Although NASA's charter and practice have been to facilitate private space businesses, the fruition of this industry has taken the policymakers by surprise. To some extent the massive growth in satellite communications and related applications was foreseen and nurtured by a combination of industry and agency efforts. Even here, however, the extent of the boom was not foreseen. The FCC, for instance, had to place a moratorium on applications for new licenses for satellite use.

The interest in private launch vehicles also surprised many space veterans, used to thinking in terms of the early days of space flight. On the other hand, the Reagan and Carter administrations have met with much less success in their efforts to privatize remote sensing (especially weather) satellites. New uses in manufacturing and production are similarly encouraged by NASA and other government agencies. This phenomenon, however, is not far enough along to determine the success of these efforts.

This industrial push has been in response to ideology and technology, but until the middle 1980s with very little concern for the policy and administrative issues involved. The success of this policy of encouraging private industry, however, has found the policymakers with the tremendous dilemma: How will this new business be regulated?

Domestic Rules

In addition to the international and national regulation of space efforts, the U.S. government must also consider how seemingly unrelated statutes and rules apply to space industry. Among these statutes are

1. The antitrust laws
2. Security regulations—state and federal
3. Patent laws
4. OSHA, Environmental Protection Agency (EPA), and Food and Drug Administration regulations
5. Others yet to be identified.

Antitrust Laws. These laws are important to space commerce for two reasons: the great sums of money often required and the foreign competition, which can often receive subsidies from its own country. The antitrust laws were drafted in a bygone era; to the extent that conditions have changed, those laws should no longer apply in space commerce.

The international competition in high technology, especially computers and semiconductors, may be the premiere case. Companies have pooled resources in consortia to develop the new products to compete with Japanese and European offerings. The Federal Trade Commission (FTC) has approved Semiconductor Research Cooperative (SRC) and the Microelectronics and Computer Technology Corporation (MCC) under the antitrust laws. SRC is a nonprofit organization; MCC is for profit, but it will license its patents broadly to avoid antitrust concerns.[10]

Yet the legal status of these consortia could be overturned by later administrations or by a court ruling. A specific amendment exempting speculative enterprises subject to foreign competition from the antitrust provision would reduce the risk for such enterprises.

Security Regulations. These rules also prevent domestic companies from acquiring the capital needed for initiating a new space effort. Stocks and other equity securities can be important methods of acquiring investment capital, yet the federal and many state Securities Exchange Commissions (SECs) do not look favorably on highly speculative stock offerings. Several such offers attempted on space projects have been unable to obtain the approval of SECs. As the space efforts become less speculative, however, these companies will find this problem less of an encumbrance.

Patent Laws. Patent law raises different issues affecting the inability of effective private space efforts. In this area NASA has done a good job of ameliorating possible antibusiness implications of the written law. Under Section 305 of the Space Act,[11] the burden seems to be placed on the inventor working under NASA contract to negotiate retention of the patent. NASA, however, has issued regulations, at the industry's request, to assure private companies that the inventors will not lose the rights to their inventions except under extraordinary situations.[12] (See the Appendixes F and G.)

NASA. In the interim NASA has been the agency that has nurtured and regulated private enterprise in space. In addition to its pro-development patent policy, it has maintained a similarly favorable policy for the risk insurance of its clientele. NASA requires its users to be insured. Since 1979 NASA has been permitted by law to provide insurance as negotiated between NASA and the user.[13]

NASA has two major devices that it uses to cooperate with private companies, the Joint Endeavor Agreement (JEA) and the Technical Exchange Agreement (TEA). The TEA permits a company to receive technical data from NASA. Often the company will be able to use NASA facilities, but the company pays to use the facilities. NASA receives, in return, results of the company's work as well as furthering its goal of introducing space processes to private concerns.

The more serious commitment by NASA and the private company is the JEA. Here, NASA provides several free flights on the Shuttle to test out processes. The company retains proprietary rights to its discovery, but NASA requires that the results are used by the company or released for public inspection or use. The best example is the JEA between NASA and McDonnell Douglas and Johnson and Johnson that permitted seven free flights to test out the continuous flow electropheresis for manufacturing pharmaceuticals in space.[14]

Regulations. As companies begin to do more work in space, the question arises as to which domestic regulation should and can reach activities in space. A threshold question is whether jurisdiction extends to extraterritorial activity by U.S. concerns. In another paper, Michael Kennedy and I have argued that private industry in space will raise a host of regulatory and legal questions that have not been raised before.[15] Here, I will mention several briefly: how will the FDA (Food and Drug Administration) deal with pharmaceuticals manufactured in space? Will OSHA provide safety standards, including exposure to radiation, for private workers in space? The potential for bureaucratic overlap in regulating space is one of the reasons that I suggest a Department of Space to coordinate and facilitate future space activity.

Space endeavors will undoubtedly continue to raise new problems for the regulators as we continue to find new products and uses in space. The challenge to the regulators will be to guard the rights of people on earth and in space while facilitating or at least not standing in the way of progress.

INTERNATIONAL SPACE LAW AND PRIVATE ENTERPRISE

Since the beginning of the space age, the U.S. government has pursued a regime in international law in which private enterprise would have a legitimate position in space and in the other transnational areas of the oceans and Antarctica. The United States still has the leading civilian space program (the largest and most advanced) and both the nation and corporations in it have much to gain from an international regime that permits a space venture to retain most of the benefits of its labor.

The coordination between domestic and international policy is another problem for U.S. space planners. For instance, the Intelsat agreement binds its members not to harm the viability of the organization. Intelsat has approved some regional satellite systems (e.g., Palapa) which replace noncompetitive terrestrial systems and does not affect a major monetary competition to Intelsat. Intelsat is much more jealous of its North Atlantic communications. In 1983 two U.S. companies (Orion and International Satellite) filed requests with the FCC to open up business and other communications channels that did not compete (so they claimed) with Intelsat. Intelsat, however, has objected to these companies, and the FCC is considering these requests. The conflict between U.S. foreign policy (Intelsat is a U.S. victory and showcase of its cooperative and peaceful space intentions) and U.S. domestic policy (the deregulation and expansion of telecommunications) makes for a thorny decision with implications beyond this case. For instance, if the United States suggests a worldwide solar energy consortium or an organization like Intelsat for the Western Hemisphere (to solve the problem of scarce orbital slots), potential members will note the unreliability of the United States in its Intelsat agreements and be very leery of such arrangements.[16]

Besides using direct diplomacy, the United States has formally pursued this balancing of domestic and foreign concerns in two major forums: the International Telecommunications Union (ITU) and the United Nations Committee on the Peaceful Uses of Outer Space (UNCOPUOS). In both forums, however, the United States has found it increasingly difficult to achieve these ends. This confrontation sees the major democratic space powers—ESA, Japan, and the United States—aligned against the "Group of 77" (actually more than 100 Third World nations). The rallying cry of these Third World nations is the New International Information Order (NIIO). Their goal is to acquire information and other technology transfer from the industrialized world in order to accelerate the development of the Third World economies. The Soviet Union and the Eastern bloc often play more of a strategic game, sometimes aligning with the other space powers, at other times siding with the Third World nations on these issues.

The Third World, Eastern bloc, and some Western countries have, for instance, argued that national sovereignty would be invaded by both remote sensing and direct broadcast into another country. Both issues have resulted in total impasse, and it is difficult to see any negotiated settlement of these issues in the UNCOPUOS.

The United States is party to four space treaties that make up the bulk of space law. These treaties negotiated by UNCOPUOS create a regime that permits free enterprise in space but does restrict it indirectly with regulatory requirements. Under the 1967 Outer Space Treaty, no one can "annex" or "appropriate" space. Nations are also made liable for any activity of their nationals. Moreover, under the Liabilities Convention of 1971, the standard of liability for space-caused damage on earth is absolute or strict liability with *no* damage limits. (These treaties are excerpted in Appendixes A, B, and C.)

These provisions permit private enterprise in space, but the practical thrust of them is twofold. It forces the nation to regulate its nationals more closely in space. It raises the cost to those corporations going into space, especially in terms of safety precautions and insurance coverage.

The other two treaties—the Rescue and Return of Astronauts (1968) and the Registration Treaty (1976)—as well as other provisions of the Outer Space and Liability Treaties define the rights and restrictions on private enterprise in space. The Moon Treaty, accepted by UNCOPUOS in 1979 but rejected by the United States, would impose other restrictions on future space activities. The treaty deals primarily with the mining of the moon and other celestial bodies. Its ambiguous provisions seem to limit private mining by proclaiming that these resources are the "common heritage of mankind" and providing for an "international regime" to manage such activities. The debate raises the specter of the Law of the Sea debate, and it is a very complicated and close question whether the Treaty would so limit free enterprise. Nonetheless, no major space power had ratified the Treaty by January 1984. (Appendixes C and D.)[17]

Conclusions

Space commerce needs the same type of guarantees as any business to thrive. It needs to be able to assess and limit its risk; it needs a decent return on its investment within a reasonable time frame. The government, through NASA, has endeavored to provide such an environment in which private space companies could develop. This development serves the national interest. Space commerce furthers the U.S. national policy of world leadership in economics and politics. It is a recognition that high technology is the measure of a nation's strength as the world enters the twenty-first century and expands into the solar system.

NOTES TO CHAPTER 9

1. *Satellite Communications*, August 1981, October 1981, November 1981, November 1982.
2. *Ibid.*, May 1982, June 1982.
3. *Aviation Week and Space Technology*, July 27, 1981, March 14, 1980, June 14, 1982; conversation with Charles Chafer, Vice President of SSIA, December 1982.
4. *Aviation Week and Space Technology*, November 21, 1983.
5. "A Bill to provide Encouragement and Necessary Regulation for the Commercial Development of Space," H.R. 1001, *98th Cong., 1st sess.*, January 1983.
6. See Robert M. Kane and Allan D. Vose, *Air Transportation* (Dubuque, Iowa, Kendall Hunt, 1975) for a discussion of the changes in aviation law. These revisions included the Airmail Act of 1925 (Kelly Act); 1926 Amendments; 1928 Amendments; Air Commerce Act of 1926; Air Mail Act of 1930 (McNary-Watres); Civil Aeronautics Act of 1938; 1940 Amendments; Federal Air Post Act of 1946.
7. Nathan C. Goldman, "U.S. Government Regulation of Private Space Stations: New Challenges to Old Bureaucracies," working paper, Institute for Constructive Capitalism, The University of Texas at Austin, 1983.
8. Thomas Karas, *The New High Ground: Strategies and Weapons of Space Age Wars* (New York: Simon and Shuster, 1983).
9. Nathan C. Goldman, "Space Race: The U.S. Won the Sprint. Can We Compete in the Marathon?" working paper, Institute for Constructive Capitalism, The University of Texas at Austin, 1983.
10. *Electronics*, March 10, 1982, April 21, 1982, November 30, 1982.
11. 42 USC Section 2451 *et seq.*, 1970.
12. 14 CFR Section 1214.104(a), 1981, "NASA will not acquire rights to inventions, patents or proprietary data privately funded by a user or arising out of activities for which a user has reimbursed NASA under the policies set forth herein. Moreover, in certain instances in which the NASA Administration has determined that activities may have a significant im-

pact on the public health, safety or welfare, NASA may obtain assurances from the user that the results will be made available to the public in terms and conditions reasonable under the circumstances."

13. Paul G. Dembling, "Catastrophic Accidents: Indemnification of Contractors against Third Party Liability," *Journal of Space Law* (October 1982): 1-12; Martin Menter, "Commercial Participation in Space Activities," *Journal of Space Law* (September 1981): 35; Public Law 96-48, August 1979.

14. Gerald J. Mossinghoff, "Intellectual Property Rights in Space Ventures," *Journal of Space Law* (October 1982): 117-122; *Encouraging Business Ventures in Space Technologies* (Washington, D.C.: National Academy of Public Administration, 1983), pp. 62-63.

15. Goldman, "U.S. Government Regulation of Private Space Stations."

16. *Satellite Communications*, January 1984. Incidentally, this problem points up the need for cooperation between agencies—at the least, bringing in NASA, U.S. State Department, the FCC, and Comsat in its public emanation.

17. Carl Q. Christol, *The Modern Law of Outer Space* (New York: Pergamon Press, 1982. See, especially on the Moon Treaty, *Proceedings of the Twenty-Third Colloquium on the Law of Outer Space*, International Institute of Space Law of the International Astronautical Federation, 1980.

APPENDIXES

APPENDIX A

Treaty on Principles Governing the Activities of States in the Exploration and Use of Outer Space, Including the Moon and Other Celestial Bodies

(OCTOBER 10, 1967)

Article I

The exploration and use of outer space, including the moon and other celestial bodies, shall be carried out for the benefit and in the interests of all countries, irrespective of their degree of economic or scientific development and shall be the province of all mankind.

Article II

Outer space, including the moon and other celestial bodies, is not subject to national appropriation by claim of sovereignty, by means of use or occupation or by any other means.

Article III

States Parties to the Treaty shall carry on activities in the exploration and use of outer space, including the moon and other celestial bodies, in accordance with international law, including the Charter of the United Nations, in the interest of maintaining international peace and security and promoting international cooperation and understanding.

Article IV

States Parties to the Treaty undertake not to place in orbit around the Earth any objects carrying nuclear weapons or any other kinds of weapons of mass destruction, install such weapons on celestial bodies or station such weapons in outer space in any other manner.

Article VI

States Parties to the Treaty shall bear international responsibility for national activities in outer space, including the moon and other celestial bodies, whether such activities are carried on by governmental agencies or by non-governmental entities, and for assuring that national activities are carried out in conformity with the provisions set forth in the present Treaty. The activities of non-governmental entities in outer space, including the moon and other celestial bodies, shall require authorization and continuing supervision by the appropriate State Party to the Treaty. When activities are carried on in outer space, including the moon and other celestial bodies, by an international organization, responsibility for compliance with this Treaty shall be borne both by the international organization and by the States Parties to the Treaty participating in such organization.

Article VII

Each State Party to the Treaty that launches or procures the launching of an object into outer space, including the moon and other celestial bodies, and each State Party from whose territory or facility an object is launched, is internationally liable for damage to another State Party to the Treaty or to its natural or juridical persons by such object or its component parts on the Earth, in air space or in outer space, including the moon and other celestial bodies.

Article IX

In the exploration and use of outer space, including the moon and other celestial bodies, States Parties to the Treaty shall be guided by the principle of cooperation and mutual assistance and shall conduct all their activities in outer space, including the moon and other celestial bodies, with due regard to the corresponding interest of all other States Parties to the Treaty. States Parties to the Treaty shall pursue studies of outer space, including the moon and other celestial bodies, and conduct exploration of them so as to avoid their harmful contamination and also adverse changes in the environment of the Earth, resulting from the introduction of extraterrestrial matter and, where necessary, shall adopt appropriate measures for this purpose. If a State Party to the Treaty has reason to believe that an activity or experiment planned by it or its nationals in outer space, including the moon and other celestial bodies, would cause potentially harmful interference with activities of other State Parties in the peaceful exploration and use of outer space, including the moon and other celestial bodies, it shall undertake appropriate international consultations before proceeding with any such activity or experiment. A State Party to the Treaty which has reason to believe that an activity or experiment planned by another State Party in outer space, including the moon and other celestial bodies, would cause potentially harmful interference with activities in the peaceful exploration and use of outer space, including the moon and other celestial bodies, may request consultation concerning the activity or experiment.

Article X

In order to promote international cooperation in the exploration and use of outer space, including the moon and other celestial bodies, in conformity with the purpose of this Treaty, the States Parties to the Treaty shall consider on a basis of equality any requests by other States Parties to the Treaty to be afforded an opportunity to observe the flight of space objects launched by those States.

The nature of such an opportunity for observation and the conditions under which it could be afforded shall be determined by agreement between the States concerned.

Article XI

In order to promote international cooperation in the peaceful exploration and use of outer space, States Parties to the Treaty conducting activities in outer space, including the moon and other celestial bodies, agree to inform the Secretary-General of the United Nations as well as the public and the international scientific community, to the greatest extent feasible and practicable, of the nature, conduct, locations and results of such activities. On receiving the said information, the Secretary-General of the United Nations should be prepared to disseminate it immediately and effectively.

APPENDIX B

Convention on International Liability for Damage Caused by Space Objects

(OCTOBER 9, 1973)

Article I

For purposes of this Convention:

a. The term "damage" means loss of life, personal injury or other impairment of health; or loss of or damage to property of States or of persons, natural or juridical, or property of international intergovernmental organizations;

b. The term "launching" includes attempted launching;

c. The term "Launching State" means:

　i. A State which launches or procures the launching of a space object;

　ii. A State from whose territory or facility a space object is launched;

d. The term "space object" includes component parts of a space object as well as its launch vehicle and parts thereof.

Article II

A launching State shall be absolutely liable to pay compensation for damage caused by its space object on the surface of the earth or to aircraft in flight.

Article III

In the event of damage being caused elsewhere than on the surface of the earth to a space object of one launching State or to persons or property on board such a space object by a space object of another launching State, the latter shall be liable only if the damage is due to its fault or the fault of persons for whom it is responsible.

Article IV

1. In the event of damage being caused elsewhere than on the surface of the earth to a space object of one launching State or to persons or property on board such a space object by a space object of another launching State, and of damage thereby being caused to a third State or to its natural or juridical persons, the first two States shall be jointly and severally liable to the third State, to the extent indicated by the following:

   a. If the damage has been caused to the third State on the surface of the earth or to aircraft in flight, their liability to the third State shall be absolute;

   b. If the damage has been caused to a space object of the third State or to persons or property on board that space object elsewhere than on the surface of the earth, their liability to the third State shall be based on the fault of either of the first two States or on the fault of persons for whom either is responsible.

Article VI

1. Subject to the provisions of paragraph 2 of this article, exoneration from absolute liability shall be granted to the extent that a launching State establishes that the damage has resulted either wholly or partially from gross negligence or from an act or omission done with intent to cause damage on the part of a claimant State or of natural or juridical persons it represents.

2. No exoneration whatever shall be granted in cases where the damage has resulted from activities conducted by a launching State which are not in conformity with international law including, in particular, the Charter of the United Nations [1] and the Treaty on Principles Governing the Activities of States in the Exploration and Use of Outer Space, including the moon and other celestial bodies.

Article VII

The provisions of this Convention shall not apply to damage caused by a space object of a launching State to:

1. Nationals of that launching State;

2. Foreign nationals during such time as they are participating in the operation of that space object from the time of its launching or at any stage thereafter until its descent, or during such time as they are in the immediate vicinity of a planned launching or recovery area as the result of an invitation by that launching State.

Article VIII

1. A State which suffers damage, or whose natural or juridical persons suffer damage, may present to a launching State a claim for compensation for such damage.

2. If the State of nationality has not presented a claim, another State may, in respect of damage sustained in its territory by any natural or juridical person, present a claim to a launching State.

3. If neither the State of nationality nor the State in whose territory the damage was sustained has presented a claim or notified its intention of presenting a claim, another State may, in respect of damage sustained by its permanent residents, present a claim to a launching State.

Article IX

A claim for compensation for damage shall be presented to a launching State through diplomatic channels. If a State does not maintain diplomatic relations with the launching State concerned, it may request another State to present its claim to that launching State or otherwise represent its interests under this Convention. It may also present its claim through the Secretary-General of the United Nations, provided the claimant State and the launching State are both Members of the United Nations.

Article X

1. A claim for compensation for damage may be presented to a launching State not later than one year following the date of the occurrence of the damage or the identification of the launching State which is liable.

2. If, however, a State does not know of the occurrence of the damage or has not been able to identify the launching State which is liable, it may present a claim within one year following the date on which it learned of the aforementioned facts; however, this period shall in no event exceed one year following the date on which the State could reasonably be expected to have learned of the facts through the exercise of due diligence.

Article XI

1. Presentation of a claim to a launching State for compensation for damage under this Convention shall not require the prior exhaustion of any local remedies which may be available to a claimant State or to natural or juridical persons it represents.

2. Nothing in this Convention shall prevent a State, or natural or juridical persons it might represent, from pursuing a claim in the courts or administrative tribunals or agencies of a launching State. A State shall not, however, be entitled to present a claim under this Convention in respect of the same damage for

which a claim is being pursued in the courts or administrative tribunals or agencies of a launching State or under another international agreement which is binding on the States concerned.

Article XII

The compensation which the launching State shall be liable to pay for damage under this Convention shall be determined in accordance with international law and the principles of justice and equity, in order to provide such reparation in respect of the damage as will restore the person, natural or juridical, State or international organization on whose behalf the claim is presented to the condition which would have existed if the damage had not occurred.

Article XIV

If no settlement of a claim is arrived at through diplomatic negotiations as provided for in Article IX within one year from the date on which the claimant State notifies the launching State that it has submitted the documentation of its claim, the parties concerned shall establish a Claims Commission at the request of either party.

Article XV

1. The Claims Commission shall be composed of three members: one appointed by the claimant State, one appointed by the launching State and the third member, the Chairman, to be chosen by both parties jointly. Each party shall make its appointment within two months of the request for the establishment of the Claims Commission.

2. If no agreement is reached on the choice of the Chairman within four months of the request for the establishment of the Commission, either party may request the Secretary-General of the United Nations to appoint the Chairman within a further period of two months.

Article XVI

1. If one of the parties does not make its appointment within the stipulated period, the Chairman shall, at the request of the other party, constitute a single-member Claims Commission.

2. Any vacancy which may arise in the Commission for whatever reasons shall be filled by the same procedure adopted for the original appointment.

3. The Commission shall determine its own procedure.

4. The Commission shall determine the place or places where it shall sit and all other administrative matters.

5. Except in the case of decisions and awards by a single-member Commission, all decisions and awards of the Commission shall be by majority vote.

Article XVIII

The Claims Commission shall decide the merits of the claim for compensation and determine the amount of compensation payable, if any.

Article XIX

1. The Claims Commission shall act in accordance with the provisions of Article XII.

2. The decision of the Commission shall be final and binding if the parties have so agreed; otherwise the Commission shall render a final and recommendatory award, which the parties shall consider in good faith. The Commission shall state the reasons for its decision or award.

Article XXII

1. In this Convention, with the exception of Articles XXIV to XXVII, references to States shall be deemed to apply to any international intergovernmental organization which conducts space activities if the organization declares its acceptance of the rights and obligations provided for in this Convention and if a majority of the States members of the organization are States Parties to this Convention and to the Treaty on Principles Governing the Activities of States in the Exploration and Use of Outer Space, including the Moon and other celestial bodies.

2. States members of any such organization which are States Parties to this Convention shall take all appropriate steps to ensure that the organization makes a declaration in accordance with the preceding paragraph.

3. If an international intergovernmental organization is liable for damage by virtue of the provisions of this Convention, that organization and those of its members which are States Parties to this Convention shall be jointly and severally liable.

APPENDIX C

Convention on Registration of Objects Launched into Outer Space

(SEPTEMBER 15, 1976)

Article I

For purposes of this Convention:

a. The term "Launching State" means:
   i. A State which launches or procures the launching of a space object;
   ii. A State from whose territory or facility a space object is launched;
b. The term "space object" includes component parts of a space object as well as its launch vehicles and parts thereof;
c. The term "State of Registry" means a launching State on whose registry a space object is carried in accordance with Article II.

Article II

1. When a space object is launched into earth orbit or beyond, the launching State shall register the space object by means of an entry in an appropriate registry which it shall maintain. Each launching State shall inform the Secretary-General of the United Nations of the establishment of such a registry.

2. Where there are two or more launching States in respect of any such space object, they shall jointly determine which one of them shall register the object in accordance with paragraph 1 of this article.

3. The contents of each registry and the conditions under which it is maintained shall be determined by the State of registry concerned.

Article III

1. The Secretary-General of the United Nations shall maintain a Register in which the information furnished in accordance with Article IV shall be recorded.

2. There shall be full and open access to the information in this Register.

Article IV

1. Each State of registry shall furnish to the Secretary-General of the United Nations, as soon as practicable, the following information concerning each space object carried on its registry;

   a. Name of launching State or States;

   b. A appropriate designator of the space object or its registration number;

   c. Date and territory or location of launch;

   d. Basic orbital parameters, including:

      i. Nodal period,

      ii. Inclination,

      iii. Apogee,

      iv. Perigee;

   e. General function of the space object.

2. Each State of registry may, from time to time, provide the Secretary-General of the United Nations with additional information concerning a space object carried on its registry.

3. Each State of registry shall notify the Secretary-General of the United Nations, to the greatest extent feasible and as soon as practicable, of space objects concerning which it has previously transmitted information, and which have been but no longer are in earth orbit.

APPENDIX D

Agreement Governing the Activities of States on the Moon and Other Celestial Bodies

(1979)

Article I

1. The provisions of this Agreement relating to the moon shall also apply to other celestial bodies within the solar system, other than the earth, except insofar as specific legal norms enter into force with respect to any of these celestial bodies.

2. For the purposes of this Agreement reference to the moon shall include orbits around or other trajectories to or around it.

3. This Agreement does not apply to extraterrestrial materials which reach the surface of the earth by natural means.

Article II

All activities on the moon, including its exploration and use, shall be carried out in accordance with international law, in particular, the Charter of the United Nations, and taking into account the Declaration of Principles of International Law concerning Friendly Relations and Cooperation among States in accordance with the Charter of the United Nations, adopted by the General Assembly on 24 October 1979, in the interest of maintaining international peace and security and promoting international cooperation and mutual understanding, and with due regard to the corresponding interests of all other States Parties.

Article III

1. The moon shall be used by all States Parties exclusively for peaceful purposes.

2. Any threat or use of force or any other hostile act or threat of hostile act on the moon is prohibited. It is likewise prohibited to use the moon in order to commit any such act or to engage in any such threat in relation to the earth, the moon, spacecraft, the personnel of spacecraft or man-made space objects.

3. States Parties shall not place in orbit around or other trajectory to or around the moon objects carrying nuclear weapons or any other kinds of weapons of mass destruction or place or use such weapons on or in the moon.

4. The establishment of military bases, installations and fortifications, the testing of any type of weapons and the conduct of military maneuvers on the moon shall be forbidden. The use of military personnel for scientific research or for any other peaceful purposes shall not be prohibited. The use of any equipment or facility necessary for peaceful exploration and use of the moon shall also not be prohibited.

Article IV

1. The exploration and use of the moon shall be the province of all mankind and shall be carried out for the benefit and in the interests of all countries, irrespective of their degree of economic or scientific development. Due regard shall be paid to the interests of present and future generations as well as to the need to promote higher standards of living conditions of economic and social progress and development in accordance with the Charter of the United Nations.

2. States Parties shall be guided by the principle of cooperation and mutual assistance in all their activities concerning the exploration and use of the moon. International cooperation in pursuance of this Agreement should be as wide as possible and may take place on a multilateral basis, on a bilateral basis, or through international intergovernmental organizations.

Article V

1. States Parties shall inform the Secretary-General of the United Nations as well as the public and the international scientific community, to the greatest extent feasible and practicable, of their activities concerned with the exploration and use of the moon, information on the time, purposes, locations, orbital parameters and duration shall be given in respect of each mission to the moon as soon as possible after launching, while information on the results of each mission, including scientific results, shall be furnished upon completion of the mission. In case of a mission lasting more than 60 days, information on conduct of the mission, including any scientific results shall be given periodically at 30 day intervals. For missions lasting more than six months, only significant additions to such information need be reported thereafter.

Article VI

1. There shall be freedom of scientific investigation on the moon by all States Parties without discrimination of any kind, on the basis of equality and in accordance with international law.

2. In carrying out scientific investigations and in furtherance of the provisions of this Agreement the States Parties shall have the right to collect on and remove from the moon samples of its mineral and other substances. Such samples shall remain at the disposal of those States Parties which caused them to be collected and may be used by them for scientific purposes. States Parties shall have regard to the desirability of making a portion of such samples available to other interested States Parties and the international scientific community for scientific investigation. States Parties may in the course of scientific investigations also use mineral and other substances of the moon in quantities appropriate for the support of their missions.

3. States Parties agree on the desirability of exchanging scientific and other personnel on expeditions to or installations on the moon to the greatest extent feasible and practicable.

Article VII

1. In exploring and using the moon, States Parties shall take measures to prevent the disruption of the existing balance of its environment whether by introducing adverse changes in such environment, its harmful contamination through the introduction of extraenvironmental matter or otherwise. States Parties shall also take measures to avoid harmfully affecting the environment of the earth through the introduction of extraterrestrial matter or otherwise.

2. States Parties shall inform the Secretary-General of the United Nations of the measures being adopted by them in accordance with paragraph 1 of this article and shall also to the maximum extent feasible notify him in advance of all placements by them of radioactive materials on the moon and of the purposes of such placements.

3. States Parties shall report to other States Parties and to the Secretary-General concerning areas of the moon having special scientific interest in order that, without prejudice to the rights of other States Parties, consideration may be given to the designation of such areas as international scientific preserves for which special protective arrangements are to be agreed in consultation with the competent organs of the United Nations.

Article IX

1. States Parties may establish manned and unmanned stations on the moon. A State Party establishing a station shall use only that area which is reasonable for the needs of the station and shall immediately inform the Secretary-General of the United Nations of the location and purposes of that station. Subsequently,

at annual intervals that State shall likewise inform the Secretary-General whether the station continues in use and whether its purposes have changed.

2. Stations shall be installed in such a manner that they do not impede the free access to all areas of the moon of personnel, vehicles and equipment of other States Parties conducting activities on the moon in accordance with the provisions of this Agreement or of Article I of the Treaty on Principles Governing the Activities of States in the Exploration and Use of Outer Space, including the moon and other celestial bodies.

Article X

1. States Parties shall adopt all practicable measures to safeguard the life and health of persons on the moon. For this purpose they shall regard any person on the moon as an astronaut within the meaning of Article V of the Treaty on Principles Governing the Activities of States on the Exploration and Use of Outer Space, including the moon and other celestial bodies and as part of the personnel of a spacecraft within the meaning of the Agreement on the Rescue of Astronauts, the Return of Astronauts and the Return of Objects Launched into Outer Space.

2. States Parties shall offer shelter in their stations, installations, vehicles and other facilities to persons in distress on the moon.

Article XI

1. The moon and its natural resources are the common heritage of mankind which finds its expression in the provisions of this agreement and in particular in paragraph 5 of this article.

2. The moon is not subject to national appropriation by any claim of sovereignty, by means of use or occupation, or by any other means.

3. Neither the surface nor the subsurface of the moon, nor any part thereof or natural resources in place, shall become property of any State, international, intergovernmental or non-governmental organization, national organization or non-governmental entity or of any natural person. The placement of personnel, space vehicles, equipment, facilities, stations and installations on or below the surface of the moon, including structures connected with their surface or subsurface, shall not create a right of ownership over the surface or the subsurface of the moon or any areas thereof. The foregoing provisions are without prejudice to the international regime referred to in paragraph 5 of this article.

4. States Parties have the right to exploration and use of the moon without discrimination of any kind on a basis of equality, and in accordance with international law and the terms of this Agreement.

5. States Parties to this Agreement hereby undertake to establish an international regime, including appropriate procedures, to govern the exploitation of the natural resources of the moon as such exploitation is about to become fea-

sible. This provision shall be implemented in accordance with Article XVIII of this Agreement.

6. In order to facilitate the establishment of the international regime referred to in paragraph 5 of this article, States Parties shall inform the Secretary-General of the United Nations as well as the public and the international scientific community to the greatest extent feasible and practicable of any natural resources they may discover on the moon.

7. The main purpose of the international regime to be established shall include:

   a. The orderly and safe development of the natural resources of the moon;
   b. The rational management of those resources;
   c. The expansion of opportunities in the use of those resources; and
   d. An equitable sharing by all States Parties in the benefits derived from those resources, whereby the interests and needs of the developing countries as well as the efforts of those countries which have contributed either directly or indirectly to the exploration of the moon shall be given special consideration.

8. All the activities with respect to the natural resources of the moon shall be carried out in a manner compatible with the purposes specified in paragraph 7 of this article and the provisions of Article VI, paragraph 2, of this Agreement.

Article XIII

A State Party which learns of the crash landing, forced landing or other unintended landing on the moon of a space object, or its component parts, that were not launched by it, shall promptly inform the launching State Party and the Secretary-General of the United Nations.

Article XIV

1. States Parties to this Agreement shall bear international responsibility for national activities on the moon whether such activities are carried on by governmental agencies or by non-governmental entities, and for assuring that national activities are carried out in conformity with the provisions set forth in the present Agreement. States Parties shall ensure that non-governmental entities under their jurisdiction shall engage in activities on the moon only under the authority and continuing supervision of the appropriate State Party.

Article XV

1. Each State Party may assure itself that the activities of other States Parties in the exploration and use of the moon are compatible with the provisions of this Agreement. To this end, all space vehicles, equipment, facilities, stations and installations on the moon shall be open to other States parties. Such States

Parties shall give reasonable advance notice of a projected visit, in order that appropriate consultations may be held and that maximum precautions may be taken to assure safety and to avoid interference with normal operations in the facility to be visited. In pursuance of this Article, any State Party may act on its own behalf or with the full or partial assistance of any other State Party or through appropriate international organizations in the framework of the United Nations and in accordance with the Charter.

2. A State Party which has reason to believe that another State Party is not fulfilling the obligations incumbent upon it pursuant to this Agreement or that another State Party is interfering with the rights which the former State has under this Agreement, may request consultations with that Party. A State Party receiving such a request shall enter into such consultations without delay.

3. If the consultations do not lead to a mutually acceptable settlement which has due regard for the rights and interests of all the States Parties, the parties concerned shall take all measure to settle the dispute by other peaceful means of their choice and appropriate to the circumstances and the nature of the dispute. If difficulties arise in connection with the opening of consultations or if consultations do not lead to a mutually acceptable settlement, any State Party may seek the assistance of the Secretary-General without seeking the consent of any other State Party concerned in order to resolve the controversy. A State Party which does not maintain diplomatic relations with another State Party concerned shall participate in such consultations at its choice, either itself or through another State Party, or the Secretary-General as intermediary.

Article XVII

Any State Party to this Agreement may propose amendments to the Agreement. Amendments shall enter into force for each State Party to the Agreement accepting the amendments upon their acceptance by a majority of the State Parties to the Agreement and thereafter for each remaining State Party to the Agreement on the date of acceptance by it.

Article XVIII

Ten years after the entry into force of this Agreement, the question of the review of the Agreement shall be included in the provisional agenda of the United Nations General Assembly in order to consider, in the light of past application of the Agreement, whether it requires revision. However, at any time after the Agreement has been in force for five years, the Secretary-General of the United Nations, as depository, shall, at the request of one-third of the States Parties to the Agreement and with the concurrence of the majority of the States Parties, convene a conference of the States Parties to review this Agreement. A review conference shall also consider the question of the implementation of the provisions of Article XI, paragraph 5, on the basis of the principle referred to in paragraph 1 of that Article and taking into account in particular any relevant technological developments.

APPENDIX E

Convention for the Establishment of a European Space Agency

(MAY 30, 1975)

Article II
Purpose

The purpose of the Agency shall be to provide for and to promote, for exclusively peaceful purposes, cooperation among European states in space research and technology and their space applications, with a view to their being used for scientific purposes and for operational space applications systems,

- a. by elaborating and implementing a long-term European space policy, by recommending space objectives to the Member States and by concerting the policies of the Member States with respect to other national and international organizations and institutions;
- b. by elaborating and implementing activities and programs in the space field;
- c. by coordinating the European space program and national programs and by integrating the latter progressively and as completely as possible into the European space program, in particular as regards the development of applications satellites;
- d. by elaborating and implementing the industrial policy appropriate to its program and by recommending a coherent industrial policy to the Member States.

Article III
Information and Data

1. Member States and the Agency shall facilitate the exchange of scientific and technical information pertaining to the fields of space research and technology and their space applications, provided that a Member State shall not be required to communicate any information obtained outside the Agency if it considers that such communications would be inconsistent with the interests of its own security or its own agreements with third parties, or the conditions under which such information has been obtained.

2. In carrying out its activities under Article V, the Agency shall ensure that any scientific results shall be published or otherwise made available after prior use by the scientists responsible for the experiments. The resulting reduced data shall be the property of the Agency.

3. When placing contracts or entering into agreements, the Agency shall, with regard to the resulting inventions and technical data, secure such rights as may be appropriate for the protection of its interests of those of the Member States participating in the relevant program, and of those of persons and bodies under their jurisdiction. These rights shall include in particular the rights of access, of disclosure and of use. Such inventions and technical data shall be communicated to the participating States.

4. Those inventions and technical data that are the property of the Agency shall be disclosed to the Member States and may be used for their own purposes by these Member States and by persons and bodies under their jurisdiction, free of charge.

5. The detailed rules for the application of the foregoing provisions shall be adopted by the Council, by a two-thirds majority of all Member States.

Article IV
Exchange of Persons

Member States shall facilitate the exchange of persons concerned with work within the competence of the Agency, consistent with the application to any person of their laws and regulations relating to entry into, stay in, or departure from, their territories.

Article V
Activities and Programs

1. The activities of the Agency shall include mandatory activities, in which all Member States participate, and optional activities, in which all Member States participate apart from those that formally declare themselves not interested in participating therein.

a. With respect to the mandatory activities, the Agency shall,
   i. ensure the execution of basic activities, such as education, documentation, studies of future projects and technological research work;
   ii. ensure the elaboration and execution of a scientific program including satellites and other space systems;
   iii. collect relevant information and disseminate it to Member States, draw attention to gaps and duplication, and provide advice and assistance for the harmonization of international and national programs;
   iv. maintain regular contact with the users of space techniques and keep itself informed of their requirements.
b. With respect to the optional activities, the Agency shall ensure, in accordance with the provisions of Annex III, the execution of programs which may, in particular, include
   i. the design, development, construction, launching, placing in orbit and control of satellites and other space systems;
   ii. the design, development, construction and operation of launch facilities and space transport systems.

2. In the area of space applications, the Agency may, should the occasion arise, carry out operational activities under conditions to be defined by the Council by a majority of all Member States. When so doing, the Agency shall

a. place at the disposal of the operating agencies concerned such of its own facilities as may be useful to them;
b. ensure as required, on behalf of the operating agencies concerned, the launching, placing in orbit and control of operational application satellites;
c. carry out any other activity requested by users and approved by the Council.

The cost of such operational activities shall be borne by the users concerned.

With respect to the coordination and integration of programs referred to in Article II (3), the Agency shall receive in good time from Member States information on projects relating to new space programs, facilitate consultations among the Member States, undertake any necessary evaluation and formulate appropriate rules to be adopted by the Council by a unanimous vote of all Member States. The objectives and procedures of the internationalization of programs are set out in Annex IV.

Article VI
Facilities and Services

1. For the execution of the programs entrusted to it, the Agency
   a. shall maintain the internal capability required for the preparation and supervision of its tasks and, to this end, shall establish and operate such establishments and facilities as are required for its activities;
   b. may enter into special arrangements for the execution of certain parts of its programs by, or in cooperation with, national institutions of the Member States, or for the management by the Agency itself of certain national facilities.

2. In implementing their programs, the Member States and the Agency shall endeavor to make the best use of their existing facilities and available services as a first priority, and to rationalize them; accordingly, they shall not set up new facilities or services without having first examined the possibility of using the existing means.

Article VII
Industrial Policy

1. The industrial policy which the Agency is to elaborate and apply by virtue of Article II (d) shall be designed in particular to
   a. meet the requirements of the European space program and the coordinated national space programs in a cost-effective manner;
   b. improve the world-wide competitiveness of European industry by maintaining and developing space technology and by encouraging the rationalization and development of an industrial structure appropriate to market requirements, making use in the first place of the existing industrial potential of all Member States;
   c. ensure that all Member States participate in an equitable manner, having regard to their financial contribution, in implementing the European space program and in the associated development of space technology; in particular the Agency shall, for the execution of its programs, grant preference to the fullest extent possible to industry in all Member States, which shall be given the maximum opportunity to participate in the work of technological interest undertaken for the Agency;
   d. exploit the advantages of free competitive bidding in all cases, except where this would be incompatible with other defined objectives of industrial policy.

Other objectives may be defined by the Council by a unanimous decision of all Member States.

The detailed arrangements for the attainment of these objectives shall be those set out in Annex V and in rules which shall be adopted by the Council by a two-thirds majority of all Member States and reviewed periodically.

2. For the execution of its programs, the agency shall make the maximum use of external contractors consistent with the maintenance of the internal capability referred to in Article VI.1

Article VIII
Launchers and Other Space Transport Systems

1. When defining its missions, the Agency shall take into account the launchers or other space transport systems developed within the framework of its programs, or by a Member State, or with a significant Agency contribution, and shall grant preference to their utilization for appropriate payloads if this does not present an unreasonable disadvantage compared with other launchers or space transport means available at the envisaged time, in respect of cost, reliability and mission suitability.

2. If activities or programs under Article V include the use of launchers or other space transport systems, the participant States shall, when the program in question is submitted for approval or acceptance, inform the Council of the launcher or space transport system envisaged. If, during the execution of a program, the participating States wish to use a launcher or space transport system other than the one originally adopted, the Council shall make a decision on this change in accordance with the same rules as those applied in respect of the initial approval or acceptance of the program.

Article IX
Use of Facilities, Assistance to Member States and Supply of Products

1. Provided that their use for its own activities and programs is not thereby prejudiced, the Agency shall make its facilities available, at the cost of the State concerned, to any Member State that asks to use them for its own programs. The Council shall determine, by a two-thirds majority of all Member States, the practical arrangements under which the facilities will be made available.

2. If, outside the activities and programs referred to in Article V but within the purpose of the Agency, one or more Member States wish to engage in a project, the Council may decide by a two-thirds majority of all Member States to make available the assistance of the Agency. The resulting cost to the Agency shall be met by the Member State or States concerned.

Article XIII
Financial Contributions

1. Each Member State shall contribute to the costs of the activities and program referred to in Article V.1 (a) and, in accordance with Annex II, to the com-

mon costs of the Agency, in accordance with a scale adopted by the Council, by a two-thirds majority of all Member States, either every three years at the time of the review referred to in Article XI.5 (a) (iii), or whenever the Council, by a unanimous vote of all Member States, decides to establish a new scale. The scale of contributions shall be based on the average national income of each Member State for the three latest years for which statistics are available. Nevertheless,

  a. no Member State shall be required to pay contributions in excess of twenty-five percent of the total amount of contributions assessed by the Council to meet these costs;

  b. the Council may, by a two-thirds majority of all Member States, decide in the light of any special circumstances of a Member State to reduce its contribution for a limited period. In particular, when the annual per capita income of a Member State is less than an amount to be decided by the Council by the same majority, this shall be considered as a special circumstance within the meaning of this provision.

2. Each Member State shall contribute to the costs of each optional program covered by Article V.1 (b), unless it has formally declared itself not interested in participating therein and is therefore not a participant. Unless all participating States decide otherwise, the scale of contributions to a given program shall be based on the average national income of each participating State for the three latest years for which statistics are available. This scale shall be revised either every three years or whenever the Council decides to establish a new scale in accordance with paragraph 1. However, no participating State shall, by the operation of this scale, be required to pay contributions in excess of twenty-five percent of the total amount of contributions to the program concerned. Nevertheless, the percentage contribution to be made by each participating State shall be equivalent to at least twenty-five percent of its percentage contribution established under the provisions of paragraph 1, unless all the participating States decide otherwise when adopting the program or during the execution of the program.

Article XIV
Cooperation

1. The agency may, upon decisions of the Council taken by unanimous votes of all Member States, cooperate with other international organizations and institutions and with governments, organizations and institutions of non-member States, and conclude agreements with them to this effect.

APPENDIX F

Section 305, NASA Act of 1958, as Amended Property Rights in Inventions

a. Section 305. Whenever any invention is made in the performance of any work under any contract of the Administration, and the Administration determines that:

1. the person who made the invention was employed or assigned to perform research, development or exploration work and the invention is related to the work he was employed or assigned to perform, or that it was within the scope of his employment duties, whether or not it was made during working hours or with a contribution by the government of the use of government facilities, equipment, materials, allocated funds, information proprietary to the government, or services of government employees during working hours; or

2. the person who made the invention was not employed or assigned to perform research, development or exploration work, but the invention is nevertheless related to the contract, or to the work or duties he was employed or assigned to perform, and was made during working hours, or with a contribution from the government of the sort referred to in clause (1),

such invention shall be the exclusive property of the United States, and if such invention is patentable a patent therefor shall be issued to the United States upon application made by the Administrator, unless the Administrator waives all or any part of the rights of the United States to such invention in conformity with the provisions of subsection (f) of this section.

b. Each contract entered into by the Administrator with any party for the performance of any work shall contain effective provisions under which such party shall furnish promptly to the Administrator a written report containing full and complete technical information concerning any invention, discovery, improvement or innovation which may be made in the performance of any such work.

c. No patent shall be issued to any applicant other than the administration for any invention which appears to the Commissioner of Patents to have significant utility in the conduct of aeronautical and space activities unless the applicant files with the Commissioner, with the application or within thirty days after request therefor by the Commissioners, a written statement executed under oath setting forth the full facts concerning the circumstances under which such invention was made and stating the relationship (if any) of such invention to the performance of any work under any contract of the Administration. Copies of each such statement and the application to which it relates shall be transmitted forthwith by the Commissioner to the Administrator.

APPENDIX G

NASA Patent and Data Policy for Shuttle Services Provided to Non-U.S. Government Users, 14 CFR, Section 1214.104

a. NASA will not acquire rights to inventions, patents or proprietary data privately funded by a user, or arising out of activities for which a user has reimbursed NASA under the policies set forth herein. However, in certain instances in which the NASA Administration has determined that activities may have a significant impact on the public health, safety or welfare, NASA may obtain assurances from the user that the results will be made available to the public on terms and conditions reasonable under the circumstances.

b. The user will be required to furnish NASA with sufficient information to verify peaceful purposes and to insure Shuttle safety and NASA's and the U.S. government's continued compliance with law and the government's obligations.

APPENDIX H

H. R. 3942: A Bill to Provide for Commercialization of Expendable Launch Vehicles and Associated Services, House of Representatives

(SEPTEMBER 21, 1983)

Be it enacted by the Senate and House of Representatives of the United States of America in Congress assembled, That this Act may be cited as the "Expendable Launch Vehicle Commercialization Act."

Findings and Purpose

Section 2 (a) *Findings*—The Congress finds and declares that

1. the peaceful uses of outer space continue to be of great value and to offer benefits to all mankind;
2. civilian applications of space technology have achieved a significant level of commercial and economic activity and offer the potential for growth in the future, particularly in the United States;
3. new and innovative equipment and services are being sought, created and offered by entrepreneurs in telecommunications, information services and remote sensing technology;
4. the private sector in the United States has the capability of developing and providing private satellite launching and associated services that would supplement the launching and associated services now available from the United States government;
5. the development of commercial expendable launch vehicles and associated services would enable the United States to retain its competi-

tive position vis-à-vis the same classes of foreign launch vehicles, thereby contributing to the national interest and economic well-being of the United States;

6. commercial provision of such services is not inconsistent with the national security interests of the United States; and

7. the United States should encourage and, only to the extent necessary, regulate private sector launching and associated services to provide for the national security and public safety and to carry out the obligations of the United States under international treaties affecting the use of space.

(b) *Purpose*—It is the purpose of this Act to

1. encourage the United States' private sector to provide expendable launch vehicles and associated launch services;

2. designate an agency within the executive branch that will be responsible for issuing commercial launch licenses and for ensuring the public safety and that national security interests and international obligations are met;

3. stimulate private sector applications of government-developed space technology; and

4. promote further economic growth and entrepreneurial activity in utilizing the space environment for peaceful purposes.

Definitions

Section 3. As used in this Act, the term

1. "launch" means to place, or attempt to place, a space object in a suborbital trajectory, in earth-orbit in outer space, or in non-earth orbit in outer space, by means of a launch vehicle;

2. "launch site" is the location from which the launch takes place;

3. "launch vehicle" means any system or systems constructed for the purpose of launching a space object, but does not include the payload;

4. "license" means a license issued by the United States government to authorize the launch of a non-government space object;

5. "payload" means an object which an applicant undertakes to launch, including subcomponents of the launch vehicle specifically designed or adapted for that particular payload, but excluding all other parts or components of the launch vehicle;

6. "person" means an individual or entity, other than an agency or department of the United States government and other than contractors and subcontractors acting on behalf of any such agency or department;

7. "secretary" means the Secretary of Commerce; and

8. "space object" means any object constructed for launching or operating in space, and includes component parts of such object as well as its launch vehicle and parts thereof and the payload, if any, and parts thereof.

Centralized Responsibility and Authority

Section 4 (a) The Secretary shall be responsible for carrying out the provisions of this Act, unless otherwise specified.

b. The Secretary shall designate within the Department of Commerce a primary point of contact for receiving processing and validating applications for a license under this Act. Such primary point of contact shall coordinate and facilitate all Federal actions pertinent to private sector space launches.

c. In support of non-government space launches, the Secretary shall, as appropriate, coordinate the availability of government launch property and services on an "added-cost" or lease basis and facilitate the use of government tooling and designs without seeking to recoup sunk development costs.

Licensing for Launching of Space Objects

a. Section 5. *In general*—

1. Except as provided in section 9 (b), no person may launch a space object from the territory of the United States, and no person who is a national of the United States may launch a space object from international waters or air space, except in accordance with a license issued under this section. Any person violating this subsection shall, upon conviction, be subject to a fine of up to $1,000,000 per violation and up to five years in prison or both.

2. Except for licenses issued under this Act or under the Federal Communications Act of 1934, no license, approval, waiver or exemption need be obtained from any Federal agency before launching a space object.

b. *Authority*—The Secretary shall, upon application and in accordance with the provisions of this Act, issue licenses for a launch or launches of space objects.

c. *Conditions*—The Secretary shall issue a license under subsection (b) only if the Secretary

1. receives assurances from the applicant sufficient to convince the Secretary that the applicant will meet the liability insurance requirements of Section 6 or this Act;

2. determines that the proposed launch vehicles, space objects and launch and tracking facilities satisfy public safety and national security requirements;

3. determines, in consultation with appropriate Federal agencies, that the applicant would qualify for any license, approval, waiver or exemption with respect to space object launches required to be obtained under Federal law in effect before the date of enactment of this Act, if such license, approval, waiver or exemption were still required; and

4. determines that there is reasonable assurance that the obligations of the United States under international treaties and agreements affecting outer space will continue to be met.

d. *Procedure—*

1. Any person may file with the Secretary an application for a license to launch a space object. Such application shall be filed not later than one hundred and eighty days before the first proposed launch. The application shall contain the following information:

    a. the name and address of the owner or owners of the proposed space objects;

    b. the proposed launch site or sites;

    c. the proposed trajectories of the launches and proposed orbital parameters;

    d. a description of the space objects;

    e. the proposed and reasonable alternative launch dates and times;

    f. potential safety and environmental hazards associated with any proposed launch;

    g. procedures intended to be employed for protecting the public safety with respect to the proposed launches;

    h. such minimum additional information determined to be required by the Secretary.

2. A. The applicant must notify the Secretary of any substantial changes of material fact with respect to an application when such changes are known and occur prior to the launch.

    B. Any information submitted by an applicant under this subection which is labeled as proprietary by the applicant shall not be disclosed to the public.

3. Within ninety days after the filing referred to in paragraph (1), the Secretary shall act either to issue, deny permanently or deny temporarily a license to carry out the activities requested by the applicant. The Secretary shall furnish a written report setting forth the basis of such action, including, in the event of a license denial, a discussion of alternative arrangements which would mitigate the reasons for denial.

4. Any applicant whose application is denied, or is approved with conditions unacceptable to such applicant, under paragraph (3) may file a petition with the Secretary within ten days after such action for reconsideration of such action. Within thirty days after the filing of any such petition for reconsideration, the Secretary shall hold a public hearing on the record to consider the issues raised in the petition. Petitioners shall have a reasonable opportunity at such hearing to present their views, to present evidence, including documents, depositions and oral testimony, and to examine witnesses. Within thirty days after such hearing, the Secretary shall enter a final order granting or denying the license.

5. Subsequent to a final order under paragraph (4), the petitioner may seek judicial review in the United States district court whose jurisdiction includes the location of the proposed launch site, or in the district in which the petitioner resides or may be found or is incorporated.

BIBLIOGRAPHY

Because there are hundreds of books and thousands of articles on space, this bibliography is extremely eclectic and selective. It provides only a beginning for the student or scholar entering this field of endeavor for the first time.

GENERAL

Aerospace Facts and Figures. Aerospace Industrial Association.
Agnosto, W. N. "Industrial Materials in Lunar Soil," in Richard S. Johnston, ed., *The Future United States Space Programs*, vol. 38, AAS Microfiche Series. San Diego, Calif.: American Astronautical Society, 1979, pp. 369–383.
American Astronautical Society Proceedings published by Univelt, Inc.
American Institute of Aeronautics and Astronautics. New York. Series of books and proceedings.
Blaine, B. C. D. *The End of an Era in Space Exploration.* San Diego, Calif.: American Astronautical Society, 1976.
Bluth, B. J., and S. R. McNeal, eds. *Update on Space.* Granada Hills, Calif.: National Behavior Systems, 1981.
Bova, Ben. *The High Road.* New York. Pocket Books, 1983.
Clarke, Arthur C. *The Promise of Space.* New York: Harper & Row, 1968.
Codding, George, A. Jr., and Anthony Rutkowski. *The International Telecommunications Union in a Changing World.* Dedham, Mass.: Artech House, 1982.
Collins, Michael. *Carrying the Fire.* New York: Ballantine Books, 1975.

Cooper, H. S. F. Jr. *A House in Space.* New York: Holt, Rinehart and Winston, 1976.
Criswell, D. R., and R. D. Waldon. "Utilization of Lunar Material in Space," in Richard S. Johnston, ed., *The Future of United States Space Programs,* vol. 38, AAS Microfiche Series. San Diego, Calif.: American Astronautical Society, 1979, p. 841.
Deudney, Daniel. *Space: The High Frontier in Perspective.* Worldwatch Paper 50, August 1982.
———. *Whole Earth Security: A Geopolitics of Peace.* Worldwatch Paper 55, July 1983.
Eberhart, J. "Sunsat: Collecting Solar Power in Orbit." *Science News* 113 (1979): 256-57.
Ehricke, Krafft A. "The Extraterrestrial Imperative." *Futures* 13 (1981): 107-14.
———. "Lunar Industries and Their Value for the Human Environment on Earth." *Acta Astronautica* (1974): 585-682.
———. "Space Industrial Productivity: New Options for the Future." House of Representatives, 94th Cong., 1st Sess., September 1975.
Frost & Sullivan. *The Commercial Satellite Communications Market in North America.* 1978.
———. *The NASA/DOD Space Market.* 1979.
Futures 14, no. 5 (1982), entire issue.
Gatland, Kenneth, ed. *The Illustrated Encyclopedia of Space Technology.* New York: Harmony Books, 1981.
Ginzberg, Eli, et al. *Economic Impact of Large Public Programs: The NASA Experience.* Salt Lake City, Utah: Olympus, 1976.
Glaser, Peter E. "Power from the Sun: Its Future." *Science* 162 (1968): 857-61.
Glenn, Jerome Clayton, and George S. Robinson. *Space Trek.* New York: Warner Books, 1980.
Graham, Daniel. *High Frontier: A New National Strategy.* Washington, D.C.: High Frontier and the Heritage Foundation, 1982.
Grey, Jerry. *Beachheads in Space.* New York: Macmillan, 1983.
Gump, David. *Space Processing, Products and Profits, 1983-1990.* Arlington, Va.: Space Business News, 1983.
Heiss, Klaus P. "Our R&D Economics of the Space Shuttle." *Astronautics and Aeronautics* (October 1971).
Heppenheimer, T. A. *Colonies in Space.* New York: Warner Books, 1978.
Holman, Mary A., et al. *The Political Economy of the Space Program.* Palo Alto, Calif.: Pacific, 1974.
Hopkins, Mark M. "Cost-Benefit Analysis of Space Manufacturing Facilities." Third Princeton AIAA Conference on Space Manufacturing Facilities, 1977.
James, Peter N. *Soviet Conquest from Space.* New Rochelle, N.Y.: Arlington House Publishers, 1974.

Jane's Pocket Book of Space Exploration. London: MacDonald and Jane's, 1976.

Karas, Thomas. *The New High Ground: Strategies and Weapons of Space Age Wars.* New York: Simon and Shuster, 1983.

Katz, James. "National Space Policy: The Forgotten Frontier." *Policy Study Journal* 10 (1982): 465.

Lilleand, Thomas M., and Ralph W. Kiefer. *Remote Sensing and Image Interpretation.* New York: Wiley, 1979.

Logsdon, John M. *The Decision To Go to the Moon.* Chicago: University of Chicago Press, 1970.

Martin, James. *Communications Satellite Systems.* Englewood, N.J.: Prentice-Hall, 1978.

McDougall, Walter A. "Technocracy and Statecrafts in the Space Age—Toward the History of a Saltation." *Current History* (1982): 1010-40.

National Academy of Public Administration. *Encouraging Business Ventures in Space Technology.* 1983.

Oberg, James. *Red Star in Orbit.* New York: Random House, 1982.

O'Leary, Brian. *The Fertile Stars.* New York: Everest House, 1981.

O'Neill, Gerard K. "The Colonization of Space." *Physics Today*, September 1974.

_____. *The High Frontier.* New York: William Morrow, 1977.

Office of Technology Assessment. *U.S. Civilian Space Policy.* Washington, D.C.: Government Publishing Office, April 1981.

Powers, Robert M. *The Coattails of God: The Ultimate Spaceflight—The Trip to the Stars.* New York: Warners, 1981.

Roland, Alex. "Returns to Earth." *Wilson Quarterly* 4 (1980): 83-89.

Sagan, Carl. *The Cosmic Connection.* New York: Dell, 1975.

Sandler, Todd, and William Schulze. "The Economics of Outer Space." *Natural Resources Journal* 21 (1981): 371-94.

Sheldon, Charles S. II, and Marcia S. Smith. *Space Activities of the United States, Soviet Union and Other Launching Countries/Organizations.* Congressional Research Service, Library of Congress, February 26, 1982.

Smith, Marcia S. *U.S. Civilian Space Programs 1958-1978.* Subcommittee on Space Science and Applications. U.S. House of Representatives, 97th Cong., 1st Sess., January 1981.

Soviet Space Programs: 1976-1980. Committee on Commerce, Science and Transportation, U.S. Senate, 97th Cong., 2nd Sess., December 1982.

Space Manufacturing Facilities/Space Colonies. Proceedings of Princeton Conference. New York: American Institute of Aeronautics and Astronautics, 1976.

Space Manufacturing from Non-Terrestrial Materials: The 1976 NASA Ames Study. Progress in Aeronautics and Astronautics. New York: American Institute of Aeronautics and Astronautics, 1977.

Stine, G. Harry. *The Third Industrial Revolution.* New York: Ace Books, 1975.
Vajk, J. Peter. *Doomsday Has Been Cancelled.* Culver City, Calif.: Peace Press, 1978.
Wihlborg, Clas, and Per Magnus Wijkman. "Outer Space Resources in Efficient and Equitable Use: New Frontiers for Old Principles." *Journal of Law and Economics* 24 (1981): 23-44.

JOURNALS

Aviation/Space, Journal of Aerospace Education.
Aviation Week and Space Technology.
The Commercial Space Report, Sunnyvale, Calif.
DBS News, Phillips Publishing.
L-5 Newsletter, L-5 Society.
NASA Space Activities.
Planetary Report, Planetary Society.
Satellite Communications.
Space Business News.
Space Calendar, Space Age Review, Inc.
The Space Humanization Series, Institute for the Social Science Study of Space.
Space Journal, Niagara University.
Space World, National Space Institute.

SPACE LAW

The material on space law has become quite voluminous. By 1983 well over 1,000 law review articles had been published on space law, and even the number of court cases that can be attributed to space and space commerce had begun to increase substantially.

Christol, Carl. *The Modern International Law of Outer Space.* New York: Pergamon Press, 1982.
Colloquia of the International Institute of Space Law.
Gorove, Stephen (Faculty Advisor). *Journal of Space Law.*
Gorove, Stephen, ed. *U.S. Space Law: National and International Regulations.*
Jasentuliyana, N. and R. Lee, eds. *Manual on Space Law.*
Matte, Nicholas. *Annals of Air and Space Law.*
Smith, D. D. *Space Stations: International Law and Policy.* Boulder, Colo.: Westview Press, 1979.
Space Law, Selected Basic Documents. U.S. Senate Committee on Commerce, Science and Transportation, 95th Cong., 2nd Sess., December 1979.

INDEX

Abrahamson, James, 41
Accounting firms, 119
Aeritalia, 20
Aerospace Sales, 103-105
Aerospatiale, 20, 24, 64, 65
Aetna Life Insurance Company, 59, 70, 122
AIA, 115, 116, 134
AIAA, 89
American Science and Technology Corporation, 35, 84
Antenna farms, 71, 95, 131
Antitrust Laws, 136, 138
Apter, David, 32
Arabsat, 60, 64
Ariane, 17, 27, 36, 42, 50, 51
ASAT, 4
Astro Tech International, 45, 50
AT&T, 34, 56, 58, 66, 67, 70, 116, 127, 128
Atlas Centaur, 36, 41, 42, 44, 123

Ball Aerospace, 92, 107, 112, 119
Banking, 119
Battelle Memorial, 90, 96
Bell, Daniel, 32
Bendix, 84, 85, 107, 112, 123
Boeing Co., 43, 47, 93, 106, 107, 112, 114

Booz, Allen and Hamilton, 82, 117, 121
Brazil, 4, 12, 13, 24, 61, 62, 63, 79, 80
British Aerospace, 20, 64, 65

Canada, 14, 16, 40, 61, 62, 64
Cannon, Howard, 130
Carter, James E., 84, 134, 136
Center for Space Policy, 118
Chevron Overseas Petroleum, 80
China, 4, 10, 11, 12, 13, 14, 27, 39, 62, 63, 79
CNN, 34, 68
Commerce, Overview, 15, 31-38
Commercial Cargo Spacelines, 45, 48
Communications Satellites, 14, 18, 22-23, 34, 55-72, 128
Competition and Cooperation, 9, 51-53, 127
Comsat, 17, 19, 26, 35, 58-59, 61, 66, 69-70, 84, 86, 116, 123, 127, 130
Condosat, 67
Congress, 49, 52, 58, 69, 81, 84, 116, 128, 129, 132
CFE (Electrophoresis), 90-92, 138
Cosmos, 9, 25

184 INDEX

Davis, Neil, 24
DBS, 18, 23, 50, 68-70
Dembling, Paul G., 120
Department of Commerce, 52, 53, 84, 129, 131, 132
Department of Defense, 76, 106, 115, 134
Department of Space, 133, 138
Department of Transportation, 44, 53, 116, 131, 132
Dula, Arthur, 121

Eagle Engineering, 47
Eisenhower, Dwight D., 86
ELV's, 39, 41, 42, 49, 123, 130, 132, 133
EPA, 87, 136
ERNO, 20, 42
EROS Data Center, 26, 77
ESA (Europe), 4, 9, 10, 11, 12, 85, 90-91
Eutelsat, 18, 20, 60, 64

FAA, 52, 130, 132
Fairchild Space and Electronics, 91-92, 112
FCC, 44, 52, 63, 67, 68, 69, 71-72, 116, 117, 128, 132, 133, 135, 139
Federal Express Co., 44, 45, 47, 48, 123
Force Multiplier, 4, 55
Ford Aerospace Co., 24, 63, 64, 65, 66, 67
France, 16, 17, 18, 19, 20, 34, 62, 64, 85; CNES, 17, 19, 85
Frost & Sullivan, 117
FTC, 128, 136

Gansler, Jacques, 106
GAO, 115
GAS, 90
General Dynamics, 24, 66, 67, 84, 89, 112
General Electric, 24, 64, 66, 67, 84, 89, 112
Geosat Committee, 80
Geostationary Orbit, defined, 57
Gibson, Ray, 16
Glaser, Peter, 95
Good, William, 47
Gorizant and Raduga, 26

Grumman, 91, 108, 112, 114, 115
GTE, 17, 67, 116, 128

HBO, 34, 63, 68, 117
Heiss, Klaus, 47, 48, 86
High Technology, 31-32
HLV, 50
Hudson, Gary, 46
Hughes Aircraft, 24, 50, 56, 64, 65, 66, 67, 84, 106, 108, 112, 116

IBM, 59, 70, 109, 112
IGY, 9, 21
India, 4, 10, 11, 12, 13, 14, 22, 61-62, 65, 85
Indonesia, 22, 61, 64
Infrastructure, Space, 37, 50-51, 98, 135
Inmarsat, 27, 60, 61, 62
Insurance, 51, 118-119
Intelsat, 4, 26, 50, 59-60, 61, 64, 139
Intersputnik, 26, 60
Italy, 16, 20, 61, 62, 79
Itek Corporation, 84
ITU, 57, 116, 139
IUS (Upper Stage), 36, 40, 43, 49, 50

Japan, 4, 9, 10, 11, 12, 13, 21-24, 61, 62, 65, 79, 90
JEA, Joint Endeavor Agreement, 90-92, 132, 137
Johnson and Johnson, 90-92, 96, 121, 138

Kemp, Jack, 131
Kennedy, Michael, 33, 138
Keyworth, George, 131

Landsat, 19, 34, 35, 40, 76, 79, 80, 82, 83, 84, 88, 130
Law, Space, 51-52, 55, 87, 138-140
Little, Arthur D. (Co.), 95, 117
Lockheed, 84, 106, 109, 112
L-5 Society, 97
Luton, Larry, 97

MAD, 5
Manufacturing (Materials Processing), 19, 23, 25, 35, 89-100, 132
Market Segmentation, 33, 36-37, 39-40, 78

INDEX

Marta, 20, 64
Martin, Marietta, 41, 42, 43, 44, 49, 106, 109, 112, 123
Marx, Karl, 32
MBB (Germany), 19, 24, 35, 59, 65, 86
McDonnel Douglas, 106, 109, 112, 114; CFE, 50, 90–91, 93, 121, 138; Delta, 21, 36, 41, 42; PAM, 36, 40, 43, 49
MCI, 43, 70
McKinsey Co., 49
Mesh, 64, 65
MGA (Micregravity Research Associates), 91, 96, 122
Military in Space, 4, 55–56; Expenditures (U.S.), 5; Motivation, 12–13; Soviet, 25
Mining (Celestial), 98–99
Mitsubishi Electric, 24, 42, 65
Molniya, 26
Motivations for Space, 12–15
MTV, 34, 68, 117

NAPA (Public Administration), 49, 117
NAB (Broadcasting), 69, 128
NAS (National Academy of Sciences), 87, 97
NASA, 17, 31, 41, 47, 48, 49, 56, 58, 81, 90–92, 93, 106, 115, 130–132, 137–138
NASC, 24, 134
NASDA (Japan), 21–24, 42
National Satellite Cable Association, 68, 116
New International Information Order, 139
Nippon Electric, 24
Nissan Motors, 42
NOAA, 76, 81, 129
NORAD, 52

Oak Satellite Corporation, 6
O'Neill, Gerard K., 71, 99
Orbital Sciences Corporation, 45, 49–50, 122, 123
OTV (Transfer vehicle), 95, 98, 131

Pacific American Launch Services, 45, 46

Packwood, Robert, 128
Patents, 137
Policy European, 20; Japan, 23–24; Soviet, 27; U.S., 31, 51–53, 127–141; Public, defined, 5–6
Prudential Insurance Co., 48, 122
Public/private goods, 34, 77–79, 86
Public Service Satellite Consortium, 70

Rand Corporation, 117
Ratiner, Leigh, 120–121
RCA, 24, 34, 66, 67, 84, 113, 116, 123, 128–130
Reagan, Ronald, 5, 34, 48, 53, 67, 81, 84, 91, 129, 131, 133, 134, 136
Remote Sensing, 14, 18–19, 22, 26, 34–35, 75–87, 129–130
Research and Development, 15, 48, 57, 135
Rockwell International, 42, 89, 92, 106, 110, 113, 114, 115

Salyut, 9, 25–26
SBS (Satellite Business Systems), 59, 66, 70
Science Applications, 89
SEC, 46, 137
Shearson/American Express, 122
Shuttle, 9, 19, 22, 27, 35, 36, 40–41, 42, 49, 86, 89, 91–92, 106, 131, 138
SII (Space Industries), 92, 122
Slayton, Donald "Deke," 47
Smith, Delbert D., 120
Southern Pacific Communications Corporation, 16, 67
Soviet Union (USSR), 9–10, 11, 12, 13, 25–27, 58, 60, 62
SPACE (Association), 68, 116
Space Activities Commission (Japan), 21, 23, 24
Space America, 19, 34, 84, 85, 123, 129–130
Space Lab, 9
Space lawyers, 120–121
Space Station, 35, 93, 131
Space Transportation Company, 44, 45, 47–48, 123
Spar Aerospace Ltd. (Canada), 40, 110
Aparx, 19, 35, 59, 85, 123
Spot Image (France), 19, 34, 85

186 INDEX

SPS (Solar Power), 95-98
SSIA (Space Services), 35-36, 44, 45, 46-47, 84, 122, 130
Starstruck, 45, 46
Stiennon Partners, 45, 46
St. Regis, 80
Sweden, 16, 19, 59, 62, 79, 86

TDF-1 (Europe DBS), 18, 65
TEA (Technical Exchange Agreement), 90, 132, 137
Teleconferencing, 70
Terra-mar, 82, 84
3M Corporation, 96
Titan (U.S. ELV), 27, 36, 41, 42, 48, 123
TOS (See Orbital Sciences Corporation), 49-50
Toshiba, 24, 65

Transpace (Cormier), 45, 46
Transpace Carriers (Delta), 45, 122
Transponders and frequencies, 22-23, 57, 68-69, 117
Transportation, 33, 36, 39-53, 130-131
Truax (Robert) Engineering, 45, 46
TRW, 66, 106, 110, 113, 115

Udall, Mo, 131
Uncopuos, 139-140
Unispace, 27
United Kingdom, 16, 61, 62
United Technologies Corporation, 50, 106, 110, 112, 114, 123

Western Union, 17, 66, 67, 71, 117
West Germany, 16, 18, 20, 61, 62, 63
Wirth, Timothy, 128

ABOUT THE AUTHOR

Nathan C. Goldman has a B.A. in history from the University of South Carolina, a J.D. from Duke University (North Carolina Bar), and a M.A. and Ph.D. in political science from Johns Hopkins. He has written many articles on space policy and space law and is a coeditor of *Space and Society: Choices and Challenges* published by the American Astronautical Society. He is Assistant Professor of government at the University of Texas at Austin.